Microfluidics and Nanophotonics: Science and Engineering

Volume 1

This Springer book series focus on the fundamental science and engineering of fluidics and photonics on micro/nanoscales, including microfluidics, nanophotonics, plasmonics, metamaterials, photonic components, devices and systems, as well as their applications such as biochemical sensors, tunable lens, micro-spectrometers, optical manipulation and lab on a chip. It aims to introduce latest research by leading international researchers and engineers of optofluidics, bio-imaging, silicon photonics and other cutting-edge research fields. As a collection, the book series is intended for researcher, industry, undergraduate and graduate students who are interested in fundamental microfluidics and nanophotonics, state-of-art technology of photonic device and system, metamaterials and lab on a chip or other applications.

Weiming Zhu · Ai-Qun Liu

Metasurfaces: Towards Tunable and Reconfigurable Meta-devices

Springer

Weiming Zhu
School of Optoelectronics Science and Engineering
University of Electronic Science and Technology of China
Chengdu, Sichuan, China

Ai-Qun Liu
School of Electrical and Electronic Engineering
Nanyang Technological University
Singapore, Singapore

ISSN 2730-9029 ISSN 2730-9037 (electronic)
Microfluidics and Nanophotonics: Science and Engineering
ISBN 978-981-19-6924-9 ISBN 978-981-19-6925-6 (eBook)
https://doi.org/10.1007/978-981-19-6925-6

This Springer imprint is published by the registered company Springer Nature Singapore Pte Ltd.
The registered company address is: 152 Beach Road, #21-01/04 Gateway East, Singapore 189721, Singapore

Preface

Metasurfaces are developing exciting new frontier research on reconfigurable material properties with promising applications on tunable and active devices. The combination of metasurfaces and micro-systems not only uncaps the controllability limits of optical materials but also paves the way for vast applications. This book focuses on the structural reconfiguration of metasurfaces using micro-systems, which have previously been developed for tiny machines and droplet formations.

The tunable metasurfaces based on nonlinear optical materials are widely discussed elsewhere, which are not thoroughly discussed in this book.

This book covers multi-disciplinary research on tunable and reconfigurable metasurfaces, revealing their potential applications on densely integrated devices with working frequencies ranging from GHz to infrared region. Topics like MEMS metasurfaces, frequency selective surface, photonic reconfigurable metasurfaces, and microfluidic metasurfaces are just a few examples that present lively research communities within the scope of this book.

This book is intended for undergraduate and graduate students who are interested in fundamental science and technology of micro-optics and artificial materials, researchers in the field of reconfigurable and tunable metamaterials, and engineers working on tunable lenses, Lidar, beam steering devices, etc.

Chengdu, China — Weiming Zhu
Singapore — Ai-Qun Liu

Acknowledgments

This book originates from the Ph.D. works of Dr. Weiming Zhu, Dr. Wu Zhang, Dr. Qinghua Song, and Dr. Libin Yan, focusing on the twelve-year ongoing research on tunable and reconfigurable metasurfaces, which are led and fully supported by Prof. Ai-Qun Liu.

We would like to thank all the students, engineers, and colleagues from the VALENS center of Nanyang Technological University for their valuable assistance and help.

We would like to thank the staff from Springer, in particular, Dr. Mengchu Huang and Dr. Rammohan Krishnamurthy, for their help, support, and patience in the delay of manuscript submission.

We would like to thank the support from the National Natural Science Foundation of China (Grant Nos. 61975026 and 61875030). Those research works are supported directly and indirectly by the Singapore MOE Tier 3 Grant (MOE2017-T3-1-001), NRF Grant (MOH-000926), A*STAR Grant (SERC A18A5b0056), and Singapore Water Agency Grant (PUB-1804-0082).

Chengdu, China — Weiming Zhu
Singapore — Ai-Qun Liu

Contents

Chapter 1
Introduction of Metasurfaces

The optical properties of materials, i.e., permittivity, permeability, optical chirality, etc., are defined by the interactions between their compositing molecular and incident electromagnetic waves. In nature, materials have either positive permittivity, permeability, or both, as shown in Fig. 1.1. The directions of the electrical field, magnetic field, and wave vector follow the right-handed rule when light is propagating in natural materials. In 1968, Victor Veselago proposed left-handed materials (LHMs) theoretically [1], which showed that the phase velocity has the opposite direction to the Poynting vector when permeability and permittivity become simultaneously negative. However, negative permittivity and permeability only exist in the different frequency regions. In 1999, Sir Pendry proposed a split-ring resonator (SRR) to realize negative permeability in a designable manner [2–5]. Later, this concept was experimentally demonstrated by David R. Smith using SRR structures and wire elements to control magnetic and electrical resonance frequency, respectively. As a result, both permeability and permittivity are designed to be negative at one frequency for microwave incidence [6–8]. Since then, a paradigm has been defined to realize designable electromagnetic properties of artificial materials with subwavelength structures. Such artificial materials consist of subwavelength structure arrays whose functionalities are defined by their chemical composition and the subwavelength architecture, called metamaterial (from the Greek word meta, meaning "beyond"). Metamaterials have now been demonstrated with extraordinary electromagnetic properties with working frequencies ranging from microwave to optical region, which shows vast applications such as perfect lens, dispersion compensator, and super prism, just to name a few [9–14].

Metasurfaces are 2D forms of metamaterials with subwavelength structures perpendicular to the propagating directions of incident electromagnetic waves. Metasurfaces are, in most cases, planar and optically thin, which can be realized by 2D fabrication technologies. More importantly, the subwavelength structures of optical metasurfaces, also called optical antennas, can be arranged in non-periodical forms

W. Zhu and A.-Q. Liu, *Metasurfaces: Towards Tunable and Reconfigurable Meta-devices*,
Microfluidics and Nanophotonics: Science and Engineering 1,
https://doi.org/10.1007/978-981-19-6925-6_1

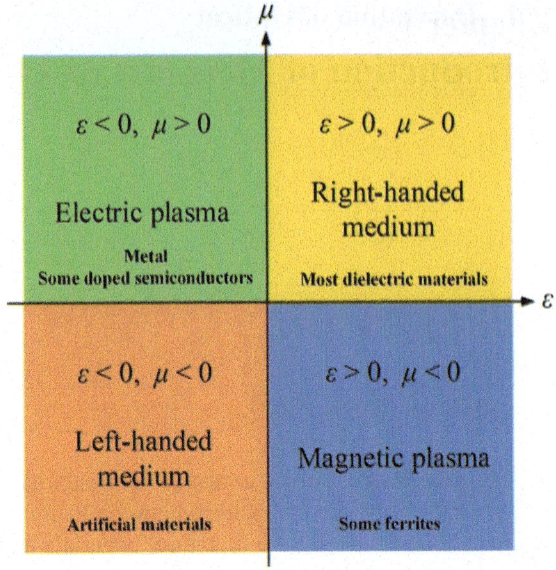

Fig. 1.1 Optical properties of materials

with a designable spatial profile of abrupt phase changes. The incident light wavefronts can be redefined by metasurfaces, which function as planar optical components or devices. The nonlinear optical effects, e.g., carrier-induced refractive indices change of the substrates or surrounding media, can tune the metasurfaces, also called active metasurfaces. The electromagnetic resonances of optical antennas can also be controlled by changing either the geometries or the arrangements of the optical antennas, which enable the metasurfaces with reconfigurable optical properties or functionalities. Now tunable and reconfigurable metasurfaces researches are significant global activities focused on the design of the functional devices with tunable, switchable capabilities, which are not only essential for metasurface based tunable/switchable devices but also compensate for the narrow band and high loss issues of the metasurfaces [15–17].

Metasurfaces have been intensively studied for decades in an electromagnetic frequency band ranging from visible, infrared, terahertz, and microwave. Vast applications based on metasurface devices have been demonstrated, such as flat metalens, high-speed beam steering, holographic display, etc. Recently, metasurfaces have been demonstrated to tailor the wavefront of the acoustic waves, which show exciting applications as acoustic insulation materials, longitudinal-wave guiding, and high-performance components for speaker devices, etc. It can be expected that metasurfaces research will continue to inspire creative designs of devices with supreme performances in terms of compact sizes, high operation speed, low costs, and less power consumption, to name a few.

1.1 Derivation of Optical Properties for Metasurfaces

Maxwell's Equations

Nature materials interact with incident electromagnetic waves with their elemental compositions, i.e., atoms and molecules. Similarly, metasurfaces respond to the incident electromagnetic waves by meta-molecules, i.e., the subwavelength unit cell structures. The incident electromagnetic waves induce the meta-molecules' surface currents or optical resonances, releasing the absorbed energy from the incident wave by coupling the near-field optical resonance mode to the propagating mode. The interactions between light and materials can be described by Maxwell's equations as follows,

Gauss's law:

$$\nabla \cdot E = \frac{\rho}{\epsilon_0} \tag{1.1}$$

Gauss's law for magnetism:

$$\nabla \cdot B = 0 \tag{1.2}$$

Faraday's law:

$$\nabla \times E = -\frac{\partial B}{\partial t} \tag{1.3}$$

Ampère's circuital law:

$$\nabla \times B = \mu_0\left(\epsilon_0 \frac{\partial \mathrm{E}}{\partial \mathrm{t}} + \mathrm{J}\right) \tag{1.4}$$

where B is the magnetic flux density, E is the electric field, ρ is the electric charge density, J is the electric current density, ϵ_0 is the permittivity of free space (8.85×10^{-12} F/m), and μ_0 is permeability of free space ($4\pi \times 10^{-7}$ H/m). Gauss's law shows that the net outward electric flux over any closed surface in free space is equal to the product of $1/\epsilon_0$ and net electric charge enclosed in the surface. Equation (1.1) is the differential form of Gauss's law. Gauss's law for magnetism can be described as Eq. (1.2), which shows that the magnetic field is a solenoidal vector field with zero divergences. The differential form of Faraday's law is shown in Eq. (1.3). Ampère's circuital law shows that time-varying electric fields can generate magnetic fields, as shown in Eq. (1.4). The relations between the electric field E, electric displacement D, magnetic field H, and the magnetic flux density B are as follows.

$$D = \epsilon_r \epsilon_0 \mathrm{E} \tag{1.5}$$

$$B = \mu_r \mu_0 H \tag{1.6}$$

where ϵ_r and μ_r are relative permittivity and permeability, respectively. The relative permittivity and permeability are defined by subwavelength structures of metasurfaces, which respond to the incident electromagnetic waves in a designable manner. The optical properties of metasurfaces can be tailored for different applications by the geometric parameters of the meta-molecules.

Effective Medium Theory

The subwavelength structures of metamaterials are designed to mimic the molecular of the natural materials by predefined geometric parameters and spatial arrangements. Metasurfaces consist of many meta-molecules that can be described by using the effective medium theory. Several methods have been introduced to retrieve the effective parameters from the simulated or experimental data of the metamaterials and metasurfaces.

The field-averaging method averages the local electric and magnetic fields to obtain the macroscopic values of the *E*, *D*, *B*, and *H* fields. The effective permittivity ϵ_{eff} and permeability μ_{eff}, which correspond to ϵ_r and μ_r of natural materials, can be derived from the averaged field values. The field-averaging method is a quick but inaccurate approach to retrieving the effective parameters, which can be applied to metamaterials and metasurfaces with deep-subwavelength meta-molecules.

The S-parameter retrieval method has long been used for antenna arrays working in the microwave region. The effective permeability and permittivity can be derived by S-parameters, which are widely used to characterize the input and output of signals of electric circuits. Both the phase and amplitude information of S-parameters are required to obtain ϵ_{eff} and μ_{eff}, which is advanced in excluding the inaccurate results by enlarging the frequency ranges of observation. However, the S-parameter retrieval method also has limitations. For example, the accurate phase information is too complicated to be measured at the frequency region above THz. It has been demonstrated in infrared frequency that ϵ_{eff} can be retrieved using the amplitude information of S-parameters only, which can be applied for limited scenarios only, e.g., when magnetic resonance is trivial ($\mu_{eff} = 1$). This method also depends on the consistency of the measured S-parameters, which might not directly quantify the optical properties of metamaterials or metasurfaces working in the high-frequency region. The S-parameter retrieval method is typically applied to the normal incidence of electromagnetic waves to simplify the simulation process, which ignores the angular dependency of meta-molecules.

The Fresnel fitting method can also be chosen to derive ϵ_{eff} and μ_{eff} of metamaterials and metasurfaces. This method had been widely applied to optical spectroscopy techniques and then introduced to characterize the optical properties of metamaterials and metasurfaces. The Fresnel equations are derived by the elastic model of optics with boundary conditions, which can be applied to oblique incident electromagnetic waves. As a result, the metasurfaces can be measured at different incident angles to derive the effective optical parameters in the absence of phase information when the angular dependency of meta-molecules is trivial. This method offers rigorous solutions to obtain the effective permeability and permittivity from the experimental data in anisotropic materials. The effective permeability and permittivity can be defined appropriately, i.e., the dimensions of meta-molecules are far smaller than that of the incident wavelength. Although it can be applied to any given incident frequency, the Fresnel fitting processes can be significantly shortened if specified functions of frequencies can describe the effective permeability and permittivity.

Figure 1.2 shows an example of deriving ϵ_{eff} and μ_{eff} by using the Fresnel fitting method. The metasurface is a split ring resonator array with oblique incidence whose electric field is parallel to the plane of the metasurface, i.e., xy-plane. The induced electric fields by incident electromagnetic fields are along the x-direction, while the induced magnetic fields are along the z-direction due to the geometries of the meta-molecules. As a result, both ϵ_{eff} and μ_{eff} can be described as 3×3 tensors by ignoring the coupling between the split ring resonators, as shown in the following.

$$\varepsilon(\omega) = \begin{pmatrix} \varepsilon_{xx}(\omega) & 0 & 0 \\ 0 & 1 & 0 \\ 0 & 0 & 1 \end{pmatrix} \tag{1.7}$$

$$\mu(\omega) = \begin{pmatrix} 1 & 0 & 0 \\ 0 & 1 & 0 \\ 0 & 0 & \mu_{zz}(\omega) \end{pmatrix} \tag{1.8}$$

The Fresnel equations can be derived from Maxwell's equations with continuous boundary conditions of both electrical and magnetic fields at the metasurface surface as follows,

$$t = \frac{e^{-ik_z d}}{\cos(q_z d) - \frac{i}{2}\left(\frac{\mu_{zz}k_z}{q_z} + \frac{q_z}{\mu_{zz}k_z}\right)\sin(q_z d)} \tag{1.9}$$

Fig. 1.2 The schematic of a split-ring-resonator metasurface with oblique incidence

$$r = \frac{\frac{i}{2}\left(-\frac{\mu_{zz}k_z}{q_z} + \frac{q_z}{\mu_{zz}k_z}\right)\sin(q_z d)}{\cos(q_z d) - \frac{i}{2}\left(\frac{\mu_{zz}k_z}{q_z} + \frac{q_z}{\mu_{zz}k_z}\right)\sin(q_z d)} \tag{1.10}$$

where t and r are transmission and reflection coefficients, respectively, q is the wave vector in medium, and k is the wave vector in vacuum. d is the thickness of the unit cell.

The electrical and magnetic response for the split ring resonator can be described by the Drude model as follows,

$$\varepsilon_{xx}(\omega) = \varepsilon_s - \frac{A_e\omega_p^2}{\omega^2 - \omega_{e0}^2 + i\omega r_e} \tag{1.11}$$

$$\mu_{zz}(\omega) = \varepsilon_s - \frac{A_m\omega^2}{\omega^2 - \omega_{m0}^2 + i\omega r_m} \tag{1.12}$$

The ε_{xx} (ω) and μ_{zz} (ω) satisfy,

$$\frac{q_y^2}{\mu_{zz}} + \frac{q_z^2}{1} = \omega^2\varepsilon_{xx} \tag{1.13}$$

The material parameters $(A_e, A_m, \omega_{m0}, \omega_{e0}, \omega_{mp}, \gamma_m, \gamma_e, \varepsilon_s)$ can be derived by submitting Eqs. 1.11 and 1.12 to Eqs. 1.9 and 1.10 and then fitting the experimental results. The Eqs. 1.10 and 1.9 are calculated using initial material parameters with reasonable guessing. Then the results are compared with the measured transmission and reflection coefficients. The simulated spectra derived by Eqs. 1.9 and 1.10 are plotted once the least-squares reach the threshold values. The Fresnel fitting method can also be accelerated by deep learning or genetic algorithms.

1.2 Designs of Meta-molecules

The meta-molecules are designed to have different geometry parameters, e.g., the sizes of gaps for split-ring resonators. As a result, the abrupt phase modulation of incident electromagnetic waves can cover or at least close to the 2-π range. The arbitrary wavefront of control of output light can be realized by arranging the meta-molecules spatial distribution with subwavelength resolutions. Therefore, the design of meta-molecules is the first step to realizing a metasurface with desired functionality. Vast design principles of meta-molecules have been proposed based on different physics principles, which are briefly introduced in this section.

Meta-molecules based on Electrical Resonances

The phase modulation of an incident electromagnetic wave by a meta-molecule can be described, similar to its counterpart of natural materials, using a mass-spring model, the phase delay of which is up to π. Therefore, meta-molecules based on electrical resonances are designed with multiple resonance modes to achieve 2-π phase modulation.

V-shaped antennas are typical designs for microwave transmitters, which have now been applied to the design of optical antennas. A V-shaped antenna is typically composed of a V-shaped metal stripe with the induced angle of θ. The single-layered V-shape metal stripe has a trivial magnetic response with normal incidence. Therefore, only electrical resonances are discussed here. The V-shaped antennas have two orthogonal resonant modes, i.e., symmetric and asymmetric modes, which are named by their spatial distributions of electric currents, as shown in Fig. 1.3. The incident electric fields are composed of two components $\boldsymbol{E_s}$ and $\boldsymbol{E_a}$, which are along the $\tilde{s}$ and $\tilde{\boldsymbol{a}}$ directions, respectively. The symmetric resonance mode is excited by $\boldsymbol{E_s}$, orientating along the symmetric axis of the V-shaped antenna. The induced electric current is symmetrically distributed on the two arms of the V-shaped antenna, which can be described by the same physics model of a straight antenna with a length of L. The effective wavelength $\lambda_{eff} = 2L$ is defined by the first-order resonance frequency of a straight antenna. The asymmetric resonant mode excited by $\boldsymbol{E_a}$ has a longer effective wavelength where $\lambda_{eff} = 4L$ since the electric current distribution is asymmetric with a longer travel distance, as shown in Fig. 1.3. In most cases, both symmetric and asymmetric modes are excited when the incident electric field is neither along $\tilde{s}$ nor $\tilde{\boldsymbol{a}}$ direction. The phase modulations of incident electromagnetic waves are different for symmetric and asymmetric modes due to the discrepancy in their resonance frequencies. Therefore, a V-shaped antenna's overall phase modulation of incidence can be tailored by changing the phase and amplitude of the two resonance modes.

The C-shaped split-ring resonator is another design of meta-molecules based on orthogonal electrical resonances. Similar to V-shaped antennas, both symmetric and asymmetric resonance modes can be excited within the C-shaped split-ring resonators, as shown in Fig. 1.4. The transmission spectrum of an arbitrary incidence has two resonance dips, as shown in Fig. 1.4a. The resonance frequencies of symmetric and asymmetric modes can be found by rotating the polarization angle of incident electromagnetic waves during the experimental characterization or numerical analysis. A linear polarized incidence is recommended for the characterization of V-shaped and C-shaped antennas since only one resonance mode is excited when the polarization direction is parallel to its axis. It should be pointed out that the efficiency of metasurfaces based on electrical resonances is very low since both the amplitude and phase modulation contribute to the overall phase of the output electromagnetic waves. It is challenging to design meta-molecules covering 2-π phase modulation while maintaining a high transmission or reflection amplitude. Therefore, two approaches have been proposed to solve this problem.

One is to use metal–insulator-metal (MIM) structures composed of a dielectric layer sandwiched by a top metasurface layer and bottom metal ground plane, as

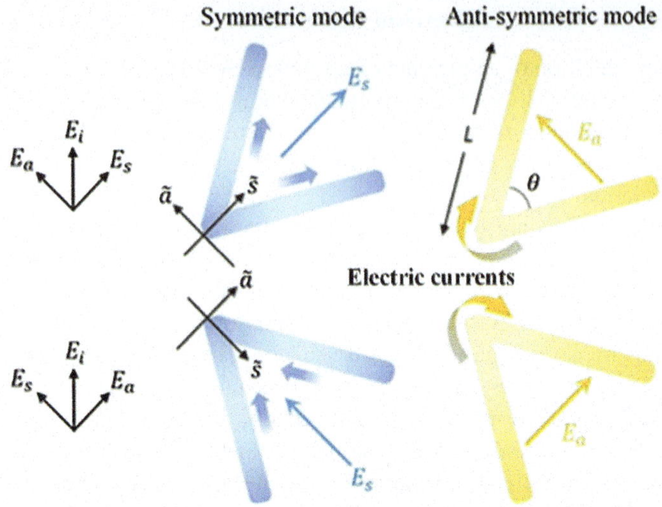

Fig. 1.3 V-shaped antenna with two orthogonal resonant modes. E_i is the incident electric field. E_s and E_a are electric-field components along with the symmetric axis $\tilde{s}$ and asymmetric axis $\tilde{a}$. L is the arm length of the V-shaped antenna. θ is the included angle of the V-shaped antenna

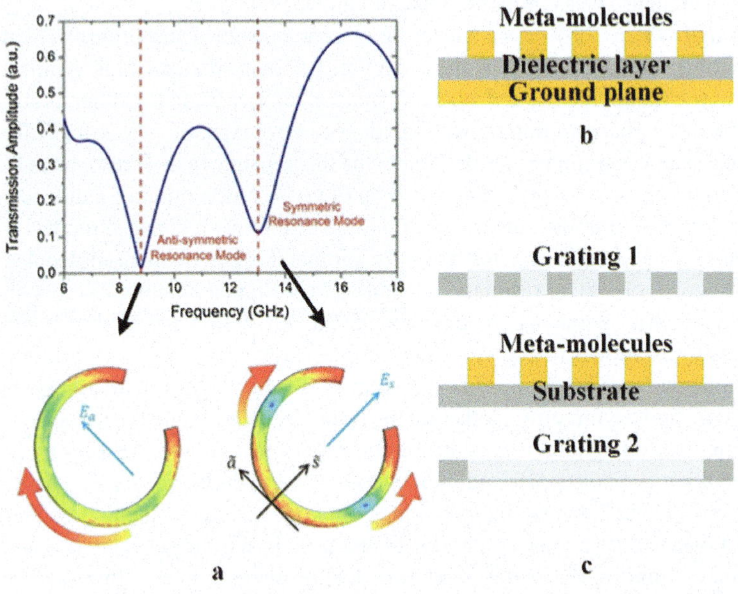

Fig. 1.4 **a** The anti-symmetric and the asymmetric resonant mode of a C-shaped split-ring resonator. **b** A reflective metasurface with the MIM structure. **c** A transmissive metasurface with two gratings

shown in Fig. 1.4b. The MIM metasurfaces work in the reflection mode, i.e., the output electromagnetic waves are reflections of the incidences. The ground metal plane is to eliminate the transmitted energy of the incidence while all the incident power is either reflected or absorbed. The high transmission can be maintained during the wavefront tuning when absorption is suppressed by rationally selected materials and the thickness of the dielectric layer.

The other method is to sandwich the metasurfaces with two orthogonally placed subwavelength gratings, as shown in Fig. 1.4c. The abrupt phase change of metasurfaces based on orthogonal resonance modes comes with 90° polarization rotation for linear polarized electromagnetic waves. The grating-assisted metasurfaces work in the transmission mode, i.e., the output electromagnetic waves are the transmissions of the incidences. The efficiencies of the metasurfaces are increased by controlling the polarization of the output waves while eliminating the reflection. Metasurfaces based on two orthogonal electric resonances also suffer from high absorptions, polarization rotation of the output waves, and high dispersions, which limit their practical applications. Despite the disadvantages mentioned above, the orthogonal resonances model is a straightforward meta-molecules design paradigm that has already been applied to many devices.

Geometric Phase Modulation of Meta-molecules

In both classical and quantum mechanics, phase modulation of the 2-π range can be achieved when a system is subjected to cyclic adiabatic processes. This phenomenon is due to the geometrical properties of the parameter space of the Hamiltonian, which is, therefore, named as geometric phase [18–20]. After being independently discovered by T. Kato in 1950, S. Pancharatnam in 1956, and H. C. Longuet-Higgins in 1958, the geometric phase was generalized by Sir Michael Berry in 1984, also known as the Pancharatnam-Berry phase. Later, the geometric phase is introduced to meta-molecules designs to achieve 2-π phase modulation by rotations of the meta-molecules.

Figure 1.5 shows an example of meta-molecules design based on geometric phase. Eight rod antennas with orientation angle β ranging from 0 to π are grouped as a super lattice of the metasurface. The rod antennas function as metamolecules for incident electromagnetic waves with circular polarization. The interaction between the incidence and rod antenna can be described by the Jones matrix as follows,

$$M_J = R(\beta)\begin{pmatrix} S_o & 0 \\ 0 & S_e \end{pmatrix} R(-\beta) \tag{1.14}$$

where S_o and S_e are the complex scattering coefficients of two perpendicular linear polarizations of the meta-molecules and R is the rotation matrix,

$$R(\beta) = \begin{pmatrix} \cos\beta & -\sin\beta \\ \sin\beta & \cos\beta \end{pmatrix} \tag{1.15}$$

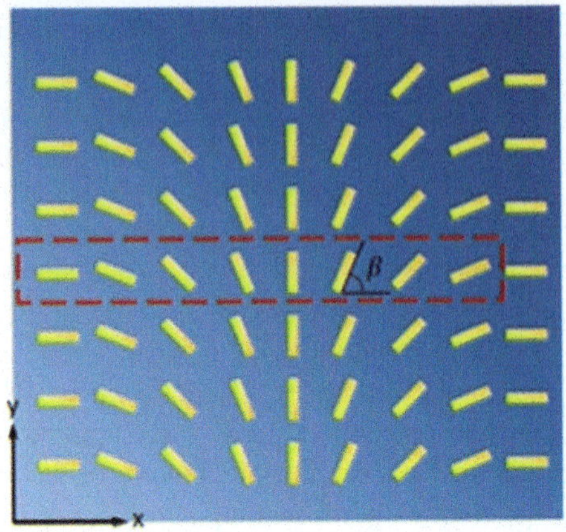

Fig. 1.5 Rod antennas array for 2 π phase modulation based on geometric phase

The meta-molecules' incident and scattered (output) electromagnetic waves are defined as E_i and E_s, respectively.

$$E_s = M_J \cdot E_i \tag{1.16}$$

Assuming the incident light has right-hand polarization and substituting Eqs. 1.16 and 1.15 to Eq. 1.14,

$$E_s = \frac{S_o + S_e}{2} E_i^R + \frac{S_o - S_e}{2} e^{-i2\beta} E_i^L \tag{1.17}$$

where E_i can be decomposed to left E_i^L and right E_i^R polarized waves and,

$$E_s = S_o e^{-i2\beta} E_i^L \tag{1.18}$$

when $S_0 = -S_e$.

As a result, the circularly polarized incidence has an abrupt phase change of 2β with polarization converted from the right-handed to the left-handed state. The meta-molecules based on the geometric phase require circularly polarized incidence and, similar to those based on orthogonal resonance modes, convert the polarization states of the incidence. Therefore, the efficiencies of the geometric metasurfaces are highly dependent on the conversion coefficient of the orthogonal polarization states. The geometric phase depends on the rotation angle β of the meta-molecules other than the frequency, which results in a nondispersive feature of phase modulation. However, the conversion coefficients of the orthogonal polarization states are still dispersive, which results in a dispersive amplitude of the output. Nevertheless, the geometric phase meta-molecules have now been widely applied to metasurfaces due to the continuous phase modulation and low dispersion.

Polarization-independent Meta-molecules

Polarization independence is required by vast applications, such as light focusing, beam steering, and electromagnetic energy harvesting, requiring meta-molecules designs to accommodate arbitrary polarization states. Many polarization-independent meta-molecules have been proposed with four-order symmetry or above. As shown in Fig. 1.6a, one example is meta-molecules composed of dielectric rods with rotational symmetry. The rod-shaped meta-molecules modulate the phase and amplitude of the incident waves based on electromagnetic resonance modes, which is polarization independent for normal incidence due to the rotational symmetry. It should be pointed out that the polarization independency cannot hold when the rod height *h* comparable with the wavelength and the incidence is oblique. An intensively studied example is two dimensional (2D) photonic crystal based on the rod-shaped unit cells where the incidence is propagating within the photonic crystal plane. The symmetry of the lattice of the meta-molecules array is another critical parameter for the design of metasurfaces considering the polarization states of the incidence, as shown in Fig. 1.6b. Although the meta-molecules have four-fold symmetry in the metasurface plane, breaking the lattice symmetry can also result in the polarization dependency of metasurfaces.

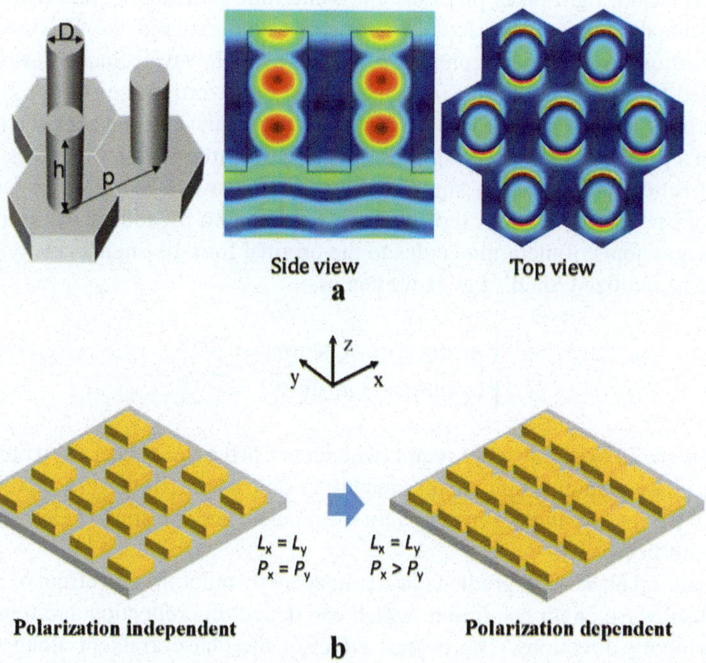

Fig. 1.6 Polarization dependency of metasurfaces. **a** One example of polarization-independent meta-molecules and **b** lattice symmetry breaking of metasurfaces

1.3 The Spatial Phase Distribution of Metasurfaces

Optical components can tailor the wavefront of the incidence based on the refraction of the electromagnetic waves, which accumulate phase differences via different optical paths. The propagation phase is proportional to the refractive index and the propagation distances. Most optically transparent materials have relatively low refractive indices, e.g., below two in visible frequencies, intensively selected to make optical components avoid high intrinsic absorptions or reflections due to the impedance mismatch. As a result, the propagation distances for refraction-based optical components must be long enough to accumulate phase differences for wavefront modulations. Therefore, refraction-based optical components are typically bulky in size and require 3D fabrication techniques in the manufacturing processes. On the other hand, the diffractive optical elements (DOE) have relatively low efficiencies and design flexibilities. In contrast, metasurfaces can tailor the wavefront of incidence with subwavelength spatial resolutions, which enable new paradigms for optical component designs.

Fermat's principle states that the light propagates along the path between two points of least time, which defines the refraction and reflection angles of the incident electromagnetic waves at two homogeneous interfaces, i.e., the Snell's law. On the other hand, light takes precisely the same time to traverse other paths that are infinitesimally close to the actual light path, which explains the physics principle of refraction-based optical components, e.g., lens. However, Snell's law has to be modified for nonhomogeneous interfaces with a nonzero phase shift gradient. The abrupt phase changes caused by the meta-molecules must be considered when using Fermat's principle or conservation of optical momentum at the interfaces to derive the refraction and reflection angle of the incident waves. Therefore, a generalized form of Snell's law is proposed by Professor Capasso by adding the abrupt phase change gradients of meta-molecules to the original form of Snell's law.

The generalized Snell's law is written as,

$$\begin{cases} n_t \sin(\theta_t) - n_i \sin(\theta_i) = \frac{1}{k_0}\frac{d\phi}{dx} \\ n_i \sin(\theta_r) - n_i \sin(\theta_i) = \frac{1}{k_0}\frac{d\phi}{dx} \end{cases} \tag{1.19}$$

where $\mathrm{k}_0 = 2\pi/\lambda_0$ is the wave vector of incidence in the free space; $\mathrm{d}\phi/\mathrm{d}x$ and $\mathrm{d}\phi/\mathrm{d}y$ are the phase gradient components along the x- and y-directions, respectively; n_i and n_t are the refractive indices of media for the incident and transmitted electromagnetic waves, respectively.

An abrupt phase shift gradient $\mathrm{d}\phi/\mathrm{d}x$ and $\mathrm{d}\phi/\mathrm{d}y$ tailor the wavefront of incidence with a subwavelength resolution, which can deflect the reflection and transmission into arbitrary directions. The overall effect of the phase gradient along the plane of metasurfaces allows the reflected and transmitted light beams to be deflected in an omnidirectional manner, enabling the metasurfaces with different functionalities, e.g., the light focusing and beam steering, just to name a few.

1.4 An Overview of Book Content

In this chapter, the basic concept of metasurfaces is discussed, including the material properties of metasurfaces described by effective medium theory, the design of meta-molecules, and the mechanism of wavefront control. The metasurface-based devices have now shown pronouncing advantages over their counterparts based on refraction or DOEs. On the other hand, the disadvantages of meta-devices, e.g., dispersion induced by meta-molecules resonances, are still crying for solutions for practical applications. This book focuses on the pioneer works based on structural reconfigurable metasurfaces and their potential applications, organized as follows.

This Chapter **Introduction of Metasurfaces** focuses on the fundamentals of metasurfaces. The materials properties of metasurfaces are discussed by effective medium theory, followed by a brief example. The meta-molecules based on different physics principles are introduced while the wavefront manipulation mechanisms are discussed. This chapter covers all the processes to design a metasurface with desired functionality.

Chapter 2 **MEMS Metasurfaces** This chapter focuses on reconfigurable metasurfaces based on mechanical actuation using microelectromechanical systems (MEMS). The discussion covers the working principle, actuation methods and corresponding tuning speed, the fabrication processes, and the pros and cons of such techniques for meta-devices.

Chapter 3 **Microfluidic Metasurfaces** This chapter focuses on liquid-based metasurfaces realized by microfluidic technologies. The discussion includes the definition of microfluidic metasurfaces, the limitations of microfluidic technologies, the fabrication processes, and the trade-off between flexibility and controllability for microfluidic meta-devices.

Chapter 4 **Tunable Electromagnetic Resonances with Slab-Split-Ring Meta-molecules** This chapter focuses on the design of metasurfaces by manipulating the electromagnetic resonances of the meta-molecules. An SSR meta-molecule is used as an example to show how the resonant modes of the meta-molecules affect the optical properties of the metasurface.

Chapter 5 **Tunable Optical Anisotropic Metasurfaces with Dynamic Control of In-plane Symmetry** This chapter focuses on the optical properties tuning based on the symmetry reconfiguration of metasurfaces. Some pioneer works on reconfigurable metasurfaces realized by symmetry breaking of meta-molecules and lattices are discussed.

Chapter 6 **Tunable Chiral Metasurfaces**. This chapter focuses on tunable chiral metasurfaces realized by MEMS and microfluidic metasurfaces. MEMS and microfluidic systems can effectively control meta-molecules' inversion symmetry, resulting in tunable chirality. Their potential applications are also discussed.

Chapter 7 **Tunable Absorber based on Meta-Fluidic-Materials**. This chapter focuses on a water-resonator-based tunable metasurface absorber working on the microwave frequency region. The working principle, as well as the experimental demonstration of the tunable absorber, are discussed.

Chapter 8 **Adaptive Metasurfaces for Dispersion Control** This chapter focuses on adaptive metasurfaces, which a microfluidic system can control. The discussion also includes the tuneable working frequency band and the experimental verifications.

Chapter 9 **Reconfigurable Metasurfaces for Dynamic Polarization Control** This chapter focuses on the microfluidic metasurface-based reconfigurable polarizers and their potential applications. The discussion includes the working principle, the tuning mechanism, and the experimental demonstration of the reconfigurable polarizer.

Chapter 10 **Tunable and Reconfigurable Flat Optics: An Outlook** The last chapter introduces the pioneer works on tunable flat optical devices based on structural reconfigurable metasurfaces. The discussion is focused on the design paradigm of reconfigurable wavefront control and its potential applications. A perspective outlook on tunable and reconfigurable flat optics is also included in this chapter.

References

1. Veselago VG (1968) The electrodynamics of substances with simultaneously negative values of ε AND μ. Soviet Phys Uspekhi 10(4):509–514. https://doi.org/10.1070/pu1968v010n04abeh003699
2. Pendry JB (2000) Negative refraction makes a perfect lens. Phys Rev Lett 85(18):3966–3969. https://doi.org/10.1103/PhysRevLett.85.3966
3. Pendry JB, Aubry A, Smith DR, Maier SA (2012) Transformation optics and subwavelength control of light. Science 337(6094):549–552. https://doi.org/10.1126/science.1220600
4. Pendry JB, Holden AJ, Robbins DJ, Stewart WJ (1999) Magnetism from conductors and enhanced nonlinear phenomena. IEEE Trans Microw Theory Tech 47(11):2075–2084. https://doi.org/10.1109/22.798002
5. Pendry JB, Holden AJ, Stewart WJ, Youngs I (1996) Extremely low frequency plasmons in metallic mesostructures. Phys Rev Lett 76(25):4773–4776. https://doi.org/10.1103/PhysRevLett.76.4773
6. Smith DR, Pendry JB, Wiltshire MCK (2004) Metamaterials and negative refractive index. Science 305(5685):788–792. https://doi.org/10.1126/science.1096796
7. Smith DR, Vier DC, Koschny T, Soukoulis CM (2005) Electromagnetic parameter retrieval from inhomogeneous metamaterials. Phys Rev E 71(3):036617. https://doi.org/10.1103/PhysRevE.71.036617
8. Schurig D, Mock JJ, Justice BJ, Cummer SA, Pendry JB, Starr AF, Smith DR (2006) Metamaterial electromagnetic cloak at microwave frequencies. Science 314(5801):977–980. https://doi.org/10.1126/science.1133628
9. Dastmalchi B, Tassin P, Koschny T, Soukoulis CM (2014) Strong group-velocity dispersion compensation with phase-engineered sheet metamaterials. Phys Rev B 89(11):115123. https://doi.org/10.1103/PhysRevB.89.115123
10. Engheta N, Ziolkowski RW (2006) Metamaterials: physics and engineering explorations. John Wiley & Sons
11. Kim M, Rho J (2015) Metamaterials and imaging. Nano Convergence 2(1):22. https://doi.org/10.1186/s40580-015-0053-7
12. Rosenblatt G, Orenstein M (2015) Perfect lensing by a single interface: defying loss and bandwidth limitations of metamaterials. Phys Rev Lett 115(19):195504. https://doi.org/10.1103/PhysRevLett.115.195504
13. Tsang M, Psaltis D (2008) Magnifying perfect lens and superlens design by coordinate transformation. Phys Rev B 77(3):035122. https://doi.org/10.1103/PhysRevB.77.035122

14. Zharov AA, Zharova NA, Noskov RE, Shadrivov IV, Kivshar YS (2005) Birefringent left-handed metamaterials and perfect lenses for vectorial fields. New J Phys 7:220–220. https://doi.org/10.1088/1367-2630/7/1/220
15. Li A, Singh S, Sievenpiper D (2018) Metasurfaces and their applications. Nanophotonics 7(6):989–1011. https://doi.org/10.1515/nanoph-2017-0120
16. He Q, Sun S, Zhou L (2019) Tunable/reconfigurable metasurfaces: physics and applications. Research 2019:1849272.https://doi.org/10.34133/2019/1849272
17. Hsiao H-H, Chu CH, Tsai DP (2017) Fundamentals and applications of metasurfaces. Small Methods 1(4):1600064. https://doi.org/10.1002/smtd.201600064
18. Berry MV (1984) Quantal phase factors accompanying adiabatic changes. Proc R Soc Lond A Math Phys Sci 392(1802):45–57. https://doi.org/10.1098/rspa.1984.0023
19. Lin D, Fan P, Hasman E, Brongersma Mark L (2014) Dielectric gradient metasurface optical elements. Science 345(6194):298–302. https://doi.org/10.1126/science.1253213
20. Luo W, Sun S, Xu H-X, He Q, Zhou L (2017) Transmissive ultrathin Pancharatnam-Berry metasurfaces with nearly 100% efficiency. Phys Rev Appl 7(4):044033. https://doi.org/10.1103/PhysRevApplied.7.044033

Chapter 2
MEMS Metasurfaces

Metasurfaces are 2D materials with rationally designed functionalities based on subwavelength meta-molecules, which have unique electromagnetic properties, such as abnormal refraction and reflection [1–3], perfect absorption [4–8], and super focusing [9–11]. However, due to the resonance nature of meta-molecules, the industrial applications of metasurface-based devices are still limited by their narrow working band and low efficiency. On the other hand, metasurfaces with tunable and reconfigurable features have been intensively studied due to their promising applications on tunable and switchable devices. More importantly, tunable metasurfaces with adaptable performances and working frequency bands may be a solution for the narrowband and low-efficiency problems.

2.1 Introduction of Tunable Metasurfaces

Two different paradigms have been developed for tunable metasurfaces after being demonstrated in the terahertz (THz) region in 2006 [12]. One is to tune the metasurface via nonlinear effects, e.g., free carrier injection, thermo-optical effect, and liquid crystal-based electrical tuning [13, 14]. The other one is to change the geometry of meta-molecules or lattice constant via mechanical actuation.

The first paradigm requires metasurfaces composed of nonlinear materials sensitive to external excitations. Tunable metasurfaces based on nonlinear materials are also called nonlinear metasurfaces. The tuning ranges of nonlinear metasurfaces are highly dependent on the nonlinearity of the materials, which are typically small in natural materials. As a result, high-Q micro-cavity designs are applied to the meta-molecules to amplify the external excitations, which downgrade the stabilities of meta-devices based on nonlinear metasurfaces. The difficulties in fabrication are another issue for meta-devices based on nonlinear metasurfaces. For example, the

W. Zhu and A.-Q. Liu, *Metasurfaces: Towards Tunable and Reconfigurable Meta-devices*, Microfluidics and Nanophotonics: Science and Engineering 1,
https://doi.org/10.1007/978-981-19-6925-6_2

most widely used nonlinear materials are challenging to handle and not compatible with massive fabrication processes, e.g., liquid crystals, phase change materials, and III–V semiconductors, which dramatically increase the fabrication cost of meta-devices.

On the other hand, the nonlinearities of some natural materials, e.g., liquid crystals, are due to the excitation of their molecules or the variation of their lattice structures. For example, the refractive index change of a liquid crystal is due to the rotation of its molecules via applied voltage, temperature variation, or electromagnetic signals. The thermo-optical effect of refractive index change for crystal silicon is mainly due to the variation of its lattice constant with the temperature. Inspired by nonlinear materials in nature, another paradigm for tunable and reconfigurable metasurfaces is proposed where the geometry and lattice constant of meta-molecules are controlled by mechanical actuation. Tunable metasurfaces based on meta-molecules geometry or lattice-constant variations are also called structural reconfigurable metasurfaces, the electromagnetic properties of which are realized by micro/nano structures reconfiguration with mechanical actuation. On the other hand, microelectromechanical systems (MEMS) have become a mature technology to realize densely integrated devices or micro/nano systems with mechanical actuation driven by external electric signals [15–17]. Therefore, the structural reconfigurable metasurfaces are combined with MEMS to realize meta-devices with compact size and fast tuning speed, called MEMS metasurfaces or micromachined metasurfaces. The fabrication process of MEMS metasurfaces is compatible with the state-of-the-art integrated circuit (IC) batch processing techniques.

Many pioneers work on radio frequency MEMS are, in fact, micromachined tunable metamaterials that have a cascade of electromagnetic resonators with structural reconfiguration. Some of the tunable electromagnetic resonators are also called MEMS metamaterials or left-handed materials. Later, this paradigm was applied to infrared metasurfaces with nanoelectromechanical systems (NEMS) [18]. Recently, MEMS metamaterials tuned by micromachined actuators, fluidics, or flexible substrates have been intensively studied due to the unlimited compositing materials and flexibilities of tuning, which lead to vast applications such as tunable lens, beam steering, optical switch, tunable filter, controllable focusing, and sensor, etc. This chapter addresses and categorizes recent progress on MEMS metasurfaces via the tuning mechanisms. Then the discussions on their pros and cons are given and followed by the conclusions and outlook of MEMS metasurfaces.

2.2 The Physics Model of MEMS Metasurfaces

The unique optical properties of metasurfaces are realized by the arrangement and design of meta-molecules' geometries. Metasurfaces can be designed with identical meta-atoms and periodic lattice structures, called homogeneous metasurfaces. The homogeneous metasurfaces have vast functionalities such as perfect absorption, optical power attenuation, beam splitting, localized electric field enhancement, etc.

On the other hand, metasurfaces are, in most cases, designed to have aperiodic lattices or inhomogeneous meta-molecules for wavefront control functionalities, which are named inhomogeneous metasurfaces. The optical properties of metasurfaces are defined by both the arrangement and scattering properties of meta-atoms.

Figure 2.1a shows an example of homogeneous metasurfaces driven by a dipole source (red arrow). The scattering properties of each meta-molecule are defined by polarizability α_{bare} as shown in Fig. 2.1b, while the lattice structure is shown in Fig. 2.1c. The optical polarizability of individual meta-molecule is determined by the linear response functions of resonances driven by the dipole source, as shown in Fig. 2.1a. The dipole field's (E_{drive}) amplitude and phase are shown in Fig. 2.1d, e, respectively. The far-field radiation pattern of the homogeneous metasurface can be described as $\frac{e^{ikR}}{R}M(k)\sum_{j}^{Scatters}a_{bare}E_{drive}(r)e^{ikr_j}$, while r_{j} is the position of jth meta-molecule and R is the distance between the metasurface and the detector pixels. Figure 2.1f, g show the simulation results of far-field radiation patterns without/with considering the coupling of the meta-atoms, respectively. The differences between Fig. 2.1g, f show that the coupling between meta-molecules, which the lattice-constant variation can control, can significantly change the far-field radiation of the homogeneous metasurfaces. In the past two decades, many MEMS metasurfaces have been proposed for tunable and reconfigurable meta-devices, most of which are based on tuning the scattering properties and coupling effects between the meta-molecules.

2.3 Meta-molecules Reconfiguration

Many meta-molecules designs have been proposed to realize metasurfaces with unique optical properties, such as large optical birefringence, abnormal refraction, and reflections, optical chirality, etc. Structural reconfiguration of meta-molecules has now been proven to be an effective method to tune the optical properties of meta-molecules, which are highly dependent on their unique structures.

The most pioneer works on reconfigurable meta-molecules using MEMS actuators are done by microwave researchers working on tunable electromagnetic resonators, which are modulated by tunable capacitors, tunable inductors, or varactor diodes driven by electrical signals. The output signals of the cascade electromagnetic resonators can be tuned via modulation of individual resonators using MEMS. This approach is later applied to the designs of the tunable metamaterials and metasurfaces. Meta-molecules with varactor diodes are widely used for fast tuning metasurfaces. The varactor diodes can control both the scatting properties and the mutual coupling between the meta-molecules. Varactor diode's effective capacitors and inductors can be modulated by external voltages or electromagnetic waves carrying the control signals. As a result, the scattering properties of meta-molecules can be modulated by varactor diodes placed inside, the working principles of which are similar to the nonlinear metasurfaces. Although the nonlinear response of varactor diodes can be designed, its tuning range is still limited by the compositing materials. Another

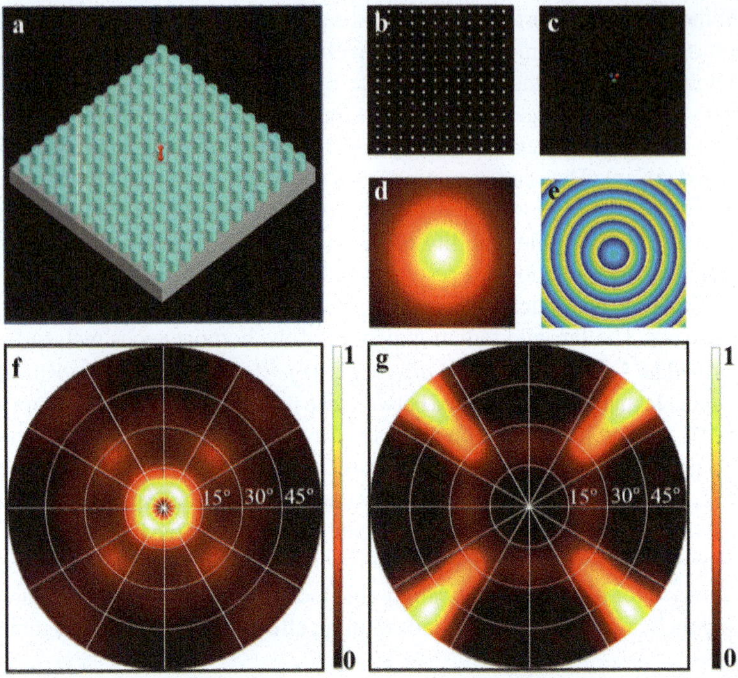

Fig. 2.1 The physics explanation of metasurface optical properties defined by meta-molecules geometry and lattice constant. **a** The schematic of a metasurface antenna. **b** and **c** are the scattering properties and lattice arrangements of meta-atoms. **d** and **e** are the amplitude and phase distribution of incident electromagnetic waves. **f** and **g** are the simulation results of far-field scattering patterns with and without considering the mutual coupling of the meta-molecules

issue is that the effective capacitors and inductors of varactor diodes vary with the driven frequency signal, which dramatically increases the design difficulties of the meta-devices, especially for those with a driven frequency higher than MHz.

Reconfigurable Electrical and Magnetic Resonances

Mechanically reconfigurable meta-molecules have been intensively studied due to the simplicity of the tuning system, i.e., micromachined actuators. MEMS technology can be applied to control the geometries of the meta-molecules. Here, a split ring resonator (SRR) is used as an example to demonstrate the working principle of MEMS reconfigurable meta-molecules.

The meta-molecule consists of two semi-square split rings, one of which is actuated by a MEMS actuator while the other is fixed, as shown in Fig. 2.2a. The gap width of the SRR can be controlled by moving the semi-square ring away or toward the fixed one via mechanical actuation. Figure 2.2b shows the equivalent circuit of the SRR meta-molecules. The capacitances of the SRR are dependent on the spatial distance between movable and fixed semi-square rings, as shown in Fig. 2.2c, d.

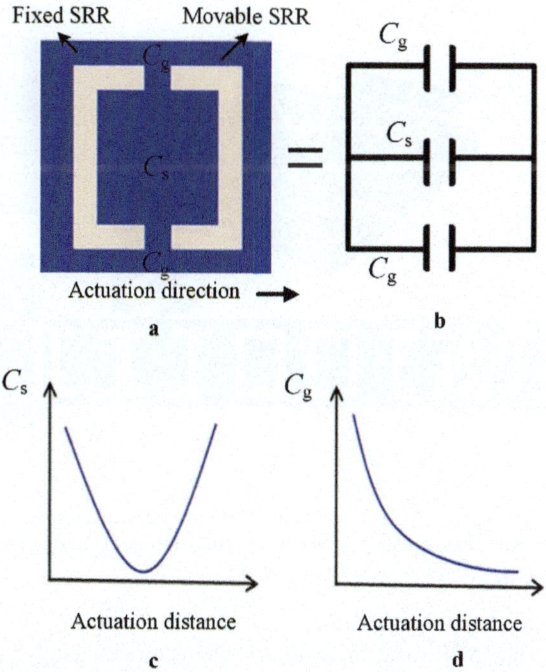

Fig. 2.2 Equivalent circuit analysis of the reconfigurable meta-molecules. **a** Shows the schematic of the meta-molecule. **b** The equivalent circuit of the meta-molecule (a). **c** and **d** show the capacitances as functions of actuation distances

The capacitances of the meta-molecule are defined as gap capacitance C_g and sidewall capacitance C_s, respectively. The gap capacitance decreases as the actuation distance increases due to the gap size between the two semi-square rings. The wall capacitance first decreases and then increases as the actuation distance increases since the meta-molecules are periodically arranged. The distances of the sidewalls are firstly increased at small actuation distance and then decreases at large actuation distances when the sidewalls of adjacent meta-molecules are closed to each other.

The meta-molecules are driven by MEMS comb drivers, as shown in Fig. 2.3a. The movable semi-squire split rings are connected with two comb drivers driven by external electrical voltage signals. The inserts show three different meta-molecules' states during the tuning processes, which are, from left to right, the middle state, the closed state, and the open state, respectively. In the middle state, the meta-molecules remain the SRR geometry shape, which can be described by the equivalent circuit as shown in Fig. 2.2b. However, in the closed state, the meta-molecules become metal rings with no gaps while, in the open state, the meta-molecules become "I" shaped metal structures. Figure 2.2b, c show the electrical field variation during the tuning processes. Moreover, Fig. 2.2d, e show the magnetic field variation during the tuning processes. Both the electrical and the magnetic resonance modes are changed during the tuning processes, which indicates a pronounced variation in the optical properties of the metasurface.

It is challenging to measure the electromagnetic response of a single meta-molecule working at THz or higher frequency region. Therefore, the tuning process

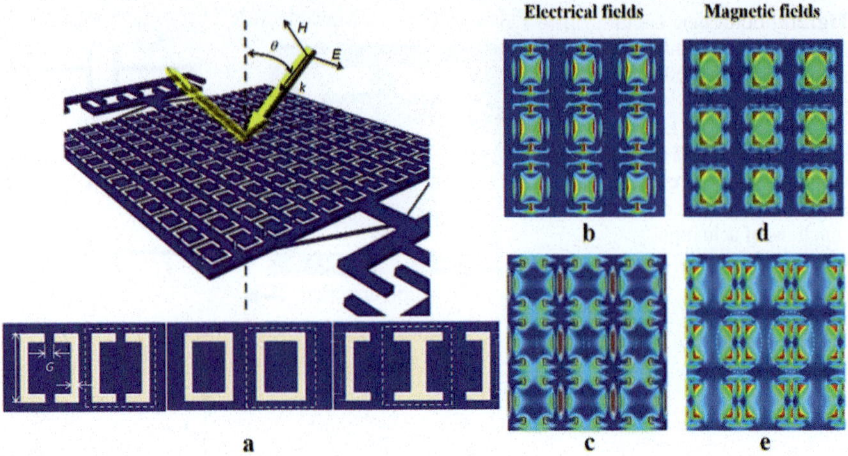

Fig. 2.3 The schematic of a MEMS metasurface with reconfigurable meta-molecules and the electromagnetic responses during the tuning process. **a** The metasurface driven by the MEMS actuator. **b** and **c** The variation of electrical fields during the tuning processes. **d** and **e** The magnetic field variation during the tuning processes

of the SRR meta-molecules is characterized using a homogenous metasurface with 400 × 400 meta-molecules. Two identical electrostatic micromachined comb actuators drive the meta-molecules array. The bidirectional translation of meta-molecules with the metasurface plane is achieved by applying electrical voltage V to the comb drive actuators where the actuation distance Δx can be approximated by $\Delta x = AV^2$. Here the actuation coefficient A is a constant defined by the design of comb-drive actuators while $A = 0.04\ \text{m/V}^2$ in this case. The semi-square SRR meta-molecules are fabricated by patterning a 0.5-μm thick aluminum (Al) layer on a single crystal silicon substrate. The SRR meta-molecules have side lengths of 19 μm, and the initial gap widths are 4 μm without external electrical voltages applied to the actuators. The Al SRR patterns have a width of 2 μm and a period of 25 μm.

In experiment, an FTIR spectrometer is used to measure the reflection spectra of the metasurface. Figure 2.4a shows the experimental setup consisting of a blackbody source, a polarizer, two diaphragms, two 50/50 beam splitters, a bolometer, a movable, and a fixed mirror. The black body radiation covers the interest frequency band of the metasurface, which ranges from 0.1 THz to 5 THz. The meta-molecules are polarization-dependent since they have only two orders of symmetry in the metasurface plane. Therefore, a linear polarizer is used to control the polarization state of the incidence. The diaphragms controlled the spot sizes of the incident and reflected THz waves, named the entrance diaphragm (for the incidence) and exit diaphragm (for reflected waves). The entrance diaphragm is to control the incident spot size to fit the size of the metasurface so that the scattering from the sample edges can be avoided. The exit diaphragm is to prevent the noise signal due to the reflection of the sample chamber. A beam splitter equally spits the radiation from the black body source. One beam is incident on the metasurface while the other serves as a reference beam with an optical path controlled by a piezo-actuated movable mirror. The reflected beam from the metasurface and the reference beam is re-joined using another 50/50 beam splitter. The interferogram is detected by a bolometer cooled by liquid helium. The air in the sample chamber is purged with Nitrogen to avoid absorption from water.

The measured reflection spectra of the metasurface with different gap widths are shown in Fig. 2.4b. The incident polarization state is transverse electric (TE) with electrical fields perpendicular to the incident plane. The incident magnetic fields have components perpendicular to the SRR structures, which have magnetic responses in the working frequency region. The open and closed states of meta-molecules are measured at $G = 8$ μm and $G = 0$ μm when the geometrical shapes of the meta-molecules correspond to the "I" and closed ring, respectively. The geometrical shapes of meta-molecules remain SRR when $G = 2$, 4, or 6 μm, which have a resonance dip induced by incident magnetic fields. The gap tuning changes both

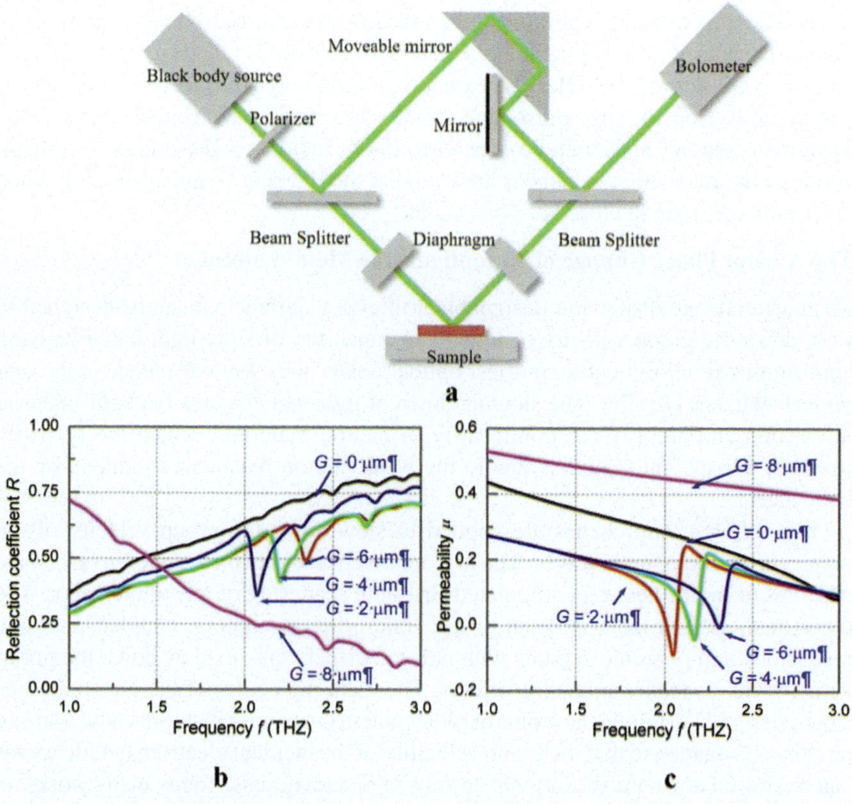

Fig. 2.4 Experimental characterization of the MEMS metasurface based on reconfigurable meta-molecules. **a** The schematics of the experimental setup, **b** the reflection spectra of the MEMS metasurface when the gap size is tuned from 0 μm to 8 μm. The different colors stand for different gap widths. **c** The spectra of effective permeability at different gaps of the two semi-square rings

magnetic resonance frequency and amplitude, which indicate a variety of effective permeability of the metasurface, as shown in Fig. 2.4b. It is worth noting that both the closed and open states of meta-molecules show no magnetic resonances as predicted by the SRR physics model introduced in Chap. 1. Therefore, the magnetic response of the metasurface can be controlled by reconfigurable meta-molecules driven by MEMS actuators.

The effective permeability of the metasurface can be approximated by using the Fresnel fitting method introduced in Chap. 1. The permeability spectra of the metasurface at different gap widths are derived by fitting the transmission and reflection spectra using a MATLAB program. The best-fitted permeabilities are chosen according to the reflection spectra with the least square values smaller than the threshold. The permeability spectra are shown in Fig. 2.4c, which shows no resonant dips and decreases as the frequency increases when $G = 0$ μm and $G = 8$ μm. The resonant dips have the same blue shift when the meta-molecules are in the middle states. The effective permeability becomes negative when the magnetic resonances are strong at frequencies 2.05 THz and 2.19 THz for $G = 2$ μm and $G = 4$ μm, respectively. The effective permeability can be tuned from negative to positive values at a given frequency. For example, the effective permeability is tuned from − 0.1 to 0.5 when G increases at 2.05 THz. The extensive tuning range of effective permeability values results from the strong resonance of magnetic fields, which only happens when meta-molecules are at the middle state with oblique incidence. For single-layer metasurfaces, the magnetic resonances are trivial at the incident frequency region when meta-molecules are in closed or open states.

The Abrupt Phase Change of Reconfigurable Meta-molecules

Metamaterials are dispersion-designable artificial materials that can be designed to slow down the group velocity of light. The capability of slow light has promising applications in all-optical memories, optical delay lines, optical regenerators, and optical switches [19–21]. The slowing down of light can enhance the light-material interaction and amplify the nonlinearity of natural materials, which are typically trivial in transparent materials due to the weak photon resonances induced by the incidence.

Two different approaches are applied to slow down the group velocity of the incident electromagnetic waves. The first approach is to rely on the photon resonance of atoms or molecules, e.g., stimulated Brillouin scattering of an atomic vapor. The other method uses meta-molecules, e.g., microring resonators, coupled photonic resonators, and photonic crystals with tailored dispersions, to slow down the group velocity via resonance tunneling effects. The meta-molecules of the metamaterials can be designed to mimic the atoms or molecular to the natural materials with a strong photonic resonance so that the group velocities of the incident electromagnetic waves can be slowed down via the rational designs of metamaterials. Today many works on slow-light metamaterials have been demonstrated with vast meta-molecules designs.

On the other hand, the metasurfaces are 2D forms of metamaterials, which only have a single or a few layers of atoms along the propagating direction of the incidence.

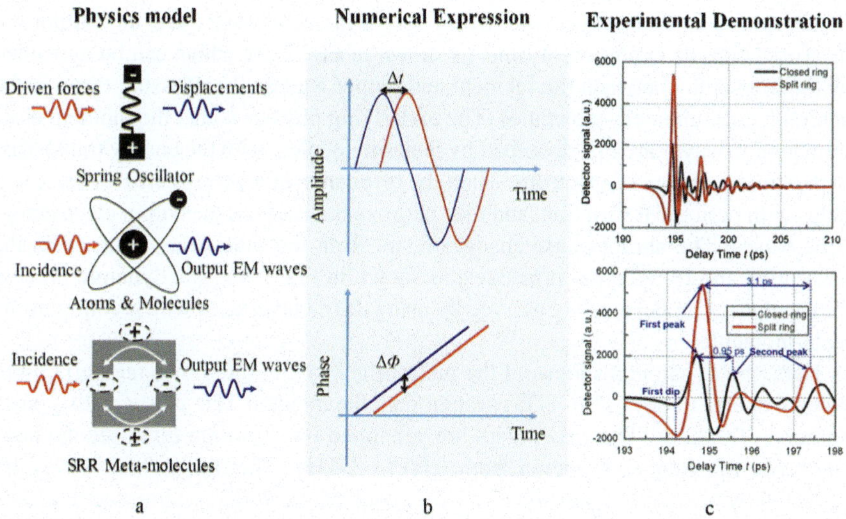

Fig. 2.5 Explanation of the abrupt phase change of meta-molecules. **a** Physics models of a spring oscillator driven by periodical forces (top), an atom excited by incident electromagnetic waves (middle), and an SRR meta-molecules (bottom). **b** The numerical expression of the time delay and phase difference of the incidence (red) and the output (blue). **c** The measured time delay of the metasurface with SRR meta-molecules. The detector's signal is a function of time for both closed state $G = 0$ (closed ring) and middle state $G = 2$ (SRR). The peaks show the times of the THz pulse resonate within the meta-molecules

Therefore, slow-light functionalities are challenging to be realized via metasurface-based devices. However, the abrupt phase change induced by meta-molecules can be explained by the photon resonances induced by the incidence, as shown in Fig. 2.5.

The abrupt phase change of metasurfaces can be explained by mechanical resonances of a spring oscillator driven by periodical forces. The displacement of the oscillator follows the direction of the driven force when the driven frequency is far lower than the intrinsic frequency of the spring oscillator. The phase delay between the driven forces and the displacements appears when the driven frequency approaches the intrinsic frequency. The maximum phase delay of a single-mode mechanical resonance can go up to π and increases to 2π with multiple resonance modes simultaneously excited by the driven forces. Resonances can also be induced in natural atoms or molecules by the incident electromagnetic waves, which apply periodical forces to the electrons and the nucleus. The displacements are now referred to as movements of the electrons. The phase shift between the incident and output electromagnetic waves is due to the resonant nature of the atoms and molecules.

Similarly, the incident electromagnetic waves can also induce electrical resonances. Take the SRR meta-molecules as an example. The electron currents are induced by the incidence on the metal parts of the semi-square structures, which results in the phase shift between the incident and output electromagnetic waves.

The phase shifts between the driven forces/incident waves and the output are invariant with the evolution of time, as shown in Fig. 2.5b, which can be explained by the time delay between the incident and output waves when passing through the natural atoms and meta-molecules. The underlying physics is that the incident electromagnetic waves are first absorbed by meta-molecules, with the energy transferred to the kinetic energy of electrons. Then the resonance of electrons creates electrical fields with periodical variation, and the energy radiates out as the output electromagnetic waves. The abrupt phase changes result from the time delay induced by the absorption and re-radiation processes, as shown in Fig. 2.5b. The dynamic process of this physics model can be observed by using the metasurfaces with reconfigurable meta-molecules.

Here, the dynamic response of the metasurface is measured by a terahertz time-domain-spectroscopy (THz-TDS) system with the incident THz pulses with a pulse duration of 0.02 ps. The THz pulses are generated by a 780-nm femtosecond laser and cover the frequency region from 0.06 to 3 THz. The output THz waves are collected by a detector controlled by a reference beam from the femtosecond laser. The detector records the output THz intensity as a function of time by varying the delay time of the reference beam, as shown in Fig. 2.5c. The red and black lines show the detector signal as the functions of time when the meta-molecules are tuned to the middle ($G = 2\ \mu m$) and closed states ($G = 0\ \mu m$), respectively.

The top graph shows the detector signal at a 20-ps period, while the bottom graph shows the zoom-in view with a 5-ps period. Due to the resonance nature of the meta-molecules, the incident THz waves are partially reflected and transmitted when reaching the boundary between the meta-molecules and the free space. As a result, the detector signal appears to be a sequence of intensity peaks. The time delay of adjacent peaks represents the effective traveling time of the THz pulse within the meta-molecules, which is inversely proportional to the group velocity. For the closed-ring meta-molecules, the traveling time is 0.95 ps corresponding to the effective refractive index of 3 when the thickness of the meta-molecules is 100 μm, which is similar to the silicon's refractive index in this frequency region. The photon resonance is weak when the meta-molecule changes to the closed state. However, in the middle state, the time delay between the first and the second peak is 3.1 ps which is two times larger than in the closed state. The group velocity of the THz pulse is slowed down to 1/3 of that in silicon. The effective refractive index is tuned from 3 to 9 when the gap between the two semi-square is changed from 0 μm to 2 μm. The tuning of the effective refractive index is more than enough for the reconfigurable abrupt phase change covering the 2-π range.

Many works have been demonstrated based on the mechanical actuated meta-molecules. In 2011, the real-time tuning of the magnetic responses with an array of 400×400 split-ring resonators was realized using the MEMS platform, which shows the effective permeability tuning from − 0.1 to 0.5 at 2.05 THz. Later, meta-molecules based on asymmetric split-ring resonators (ASRRs) are proposed to tune the dipole–dipole coupling of both magnetic and electrical fields via mechanical actuation, which shows the capability of tuning the effective permeability and permittivity simultaneously. The tuning ranges of resonance frequencies for TE and TM polarized incidence are 26% and 19% of the central frequency, respectively. Different form nonlinear meta-molecules limited by the materials properties of their compositions, the mechanically reconfigurable meta-molecules can be tuned from one geometric shape to another, e.g., from the closed state to the open state, which, therefore, results in an extensive tuning range and control flexibilities.

2.4 Lattice Reconfiguration

In nature, the nonlinear effects of crystal materials typically result from the variation of lattice structures introduced by external excitations, such as thermal expansion, electrical/optical induced phase changes, etc. For example, the lattice expansion explains the refractive index change of the crystal silicon due to the temperature change, where the spacing between the silicon atoms is inversely proportional to the free carrier density. The phase-change materials can have an extensive tuning range of material optical properties realized by the distortion of lattice structures, as shown in Fig. 2.6a. The chemical bonds between the atoms of natural crystal materials can be reformed by thermal/optical/electrical/magnetic induced external excitations. As a result, the crystal structures of phase change materials can be switched between the amorphous state, crystalline state, quasi-crystal state, etc. Similarly, metasurfaces driven by mechanical actuation have tunable optical properties via lattice reconfiguration. Furthermore, the lattice shiftings of multi-layered metasurfaces are also demonstrated by mechanical actuation, which is quite challenging to be realized using natural crystal materials, as shown in Fig. 2.6b.

In-plane Lattice Shifting

As discussed in 2.2, the electromagnetic responses of the metasurfaces are, in most cases, highly dependent on the coupling between the meta-molecules, which are dominated by lattice structures. As a result, the electromagnetic properties of metasurfaces can be tuned by changing the spacing or arrangments of the meta-molecules. Some pioneer works show that tunable metamaterials and metasurfaces can be realized by changing the coupling between the meta-molecules based on nonlinear substrates or varactor diodes. Those tunning mechanisms have limited effects on symmetry tuning of the lattice structures of the metasurfaces, which can be used to change the anisotropy states of the electromagnetic materials. Anisotropic metasurfaces and metamaterials are intensively demonstrated due to their extraordinary applications in negative refraction, super-resolution focusing, cloaking, light amplification, etc. Additionally, dichroic polarizers realized by artificial anisotropic materials have potential applications in displays and many other optical components.

The in-plane lattice reconfiguration of a MEMS metasurface is firstly demonstrated in the THz region, which shows switching between the polarization-dependent state and polarization-independent state with lattice symmetry reconfiguration. The lattice reconfiguration is realized via in-plane translation of selected meta-molecules. The lattice of the metasurface is switched between square and triangle arrangements

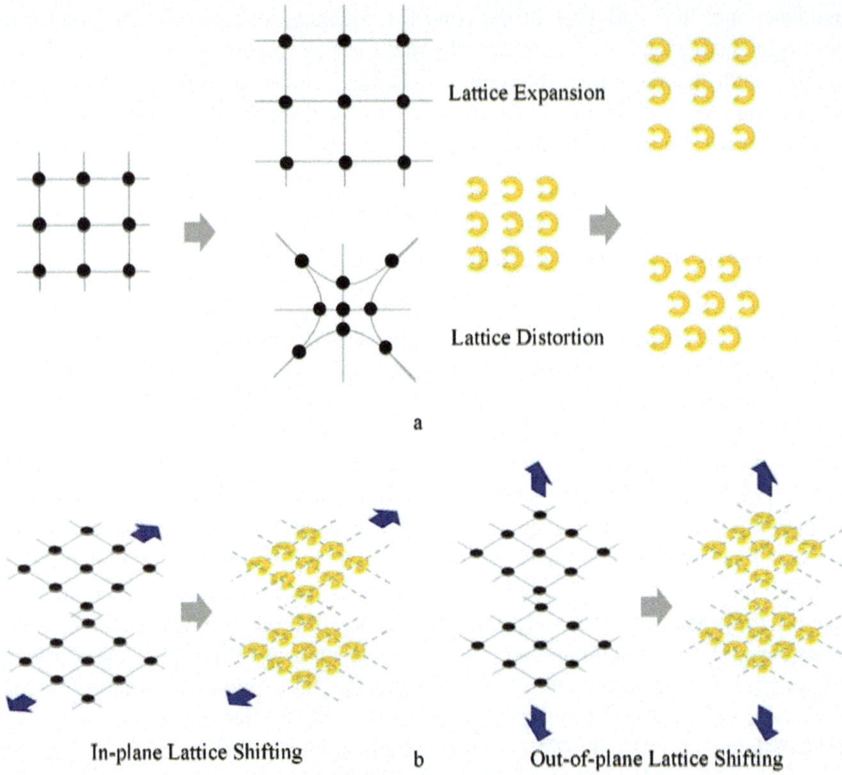

Fig. 2.6 The comparison between crystals in nature and metasurfaces. The black dots and SRRs represent natural atoms and meta-molecules, respectively. **a** The lattice expansion and distortion. **b** The in-plane and out-of-plane lattice reconfiguration for multi-layered metasurfaces

without tuning the meta-molecules. As a result, the lattice structure is switched from twofold to fourfold rotational symmetries, which are polarization-dependent and polarization-independent states, respectively. It should be pointed out that the resonance frequency of the metasurface is also changed during the lattice reconfiguration, which should be considered as the shift of the working frequency band. MEMS metasurfaces based on lattice symmetry reconfiguration show pronounced advantages compared with natural crystal materials in terms of tuning range and control flexibility, which has promising applications on photonic devices based on engineered electromagnetic anisotropy with designable and controllable manners.

Flexible electronics have now been intensively studied due to their potential applications on wearable electronic devices, biologically compatible sensors, artificial skin, etc. [22, 23]. Most of the works focus on how to realize integrated electronic circuits with a flexible substrate that can adapt to complex shapes based on desired applications. For example, an electronic circuit can be patterned on a polydimethylsiloxane (PDMS) flexible substrate, which can be bent, stretched, or deformed to any shape according to the needs. Therefore, the instability/tunability of the flexible electronics due to the flexible substrate is often considered a drawback of such devices. On the other hand, the flexible substrates are applied to the tunable and reconfigurable metasurfaces with controllable electromagnetic properties due to the lattice expansion or distortion, as shown in Fig. 2.6a. The resonance frequency of nano-photonic structures, e.g., optical antennas and gratings, can be tuned by the mechanical deformations of their flexible substrates. A straightforward demonstration is to control the dispersion angles of a grating with tunable periodicity based on a stretchable substrate. The tunable photonics based on soft-substrate metasurfaces have now been intensively reported due to their potential applications on transparent and portable/wearable photonic devices. Metasurfaces composed of soft and biocompatible materials are proposed for bio-sensors, endoscopes, and cancer elimination by laser pulses. Additionally, soft and thin films with metasurface antennas can be used to coat vehicles and aircraft for stealth functionalities. The flexible substrates can also tune the geometry of meta-molecules. The soft-substrate metasurfaces can be fabricated by using nano printing or pattern transfer technologies. The optical antennas with small feature sizes can be firstly realized by high-resolution equipment, e.g., focus ion beam (FIB) and electronic beam lithography (EBL), and then transferred to the flexible substrates. Therefore, current fabrication technologies can realize the meta-molecules working in near-infrared and visible frequency regions. Flexible metasurfaces have now become a doorway to the next generation of tunable and reconfigurable photonic devices with not only large tunability and design flexibility but also new functionalities that traditional rigid materials can never realize.

Out-of-plane Lattice Shifting

Multi-layered metamaterials/metasurfaces refer to the quasi-3D artificial materials which are realized by stacking multiple single-layered metasurfaces. The interactions between the metasurface layers enable new electromagnetic properties, such as optical clarity, near-to-zero permeability, high efficient polarization conversion, etc. Tunable metamaterials via lateral lattice reconfiguration are realized by shifting the relative positions of adjacent metasurface layers. The experiment demonstrates a continuous tuning of resonance frequency, which functions as a tunable filter. Later, the out-of-plane lattice reconfigurations of multi-layered metasurfaces are also demonstrated in the radio frequency region based on mechanical actuation. However, this technology is very difficult to apply to higher frequency regions such as infrared or visible frequency regions because the near field coupling between the adjacent layers requires subwavelength spacing of the free-hanging metasurface layers. Alternatively, some theoretical works based on metasurface enhanced optical forces or Casimir forces have been reported, some of which have prime experimental demonstrations. Nanomotors and robots based on nanostructures driven by optical forces are also intensively demonstrated due to their promising energy harvesting and nano-machine functionalities. Lattice-reconfigurable metasurfaces can be expected to enable vast applications with the development of nanomotors and robots.

Breaking the chemical bonds of the atoms and the molecules is, in most cases, a power-consuming and non-reversible process for natural materials. As a result, lattice reconfiguration is not the first choice to control their optical properties in the realization of tunable and reconfigurable photonic devices. MEMS metasurfaces have now offered a particle solution for lattice reconfiguration of artificial materials working in THz and GHz regions. More importantly, the lattice symmetry of MEMS metasurfaces can be designed and tuned in a controllable manner, which does not rely on material properties. The merits of the lattice reconfigurable metasurfaces are as follows. First, the chemical bonds do not exist between adjacent layers of metasurfaces. Second, the tunability of metasurfaces can be rationally designed by the choice of meta-molecules and lattice structures. Finally, optical anisotropy and charity can be controlled by changing the symmetry of the metasurface lattice.

2.5 Fabrication Technologies

The MEMS metasurfaces working at the THz region typically have meta-molecules with a period around tens of microns. Therefore, the minimum feature size of the fabrication is approximately several microns, which is compatible with the optical MEMS device's fabrication process based on silicon. Here, a fabrication process of SRR metasurfaces driven by MEMS actuator with electrostatic forces is discussed in this section, as shown in Fig. 2.7.

The fabrication processes can be divided into three major steps. The first step is to define the metal pattern of the meta-molecules by using ion etching or lift-off processes. The second step is to fabricate the supporting frames of the meta-molecules by using the deep reactive ion etch (DRIE) processes. The movable parts of the meta-molecules are supported by released frames, while the fixed parts are anchored on the larger supporting frames without being released. The final step is to coat metal on the sidewall of the meta-molecules where the ring resonators are split. As a result, the fixed and movable parts of the SRR resonators can be ohmically conducted at open and closed states. It is worth noting that the DRIE processes are intensively applied to fabricate MEMS metasurfaces due to the high selectivity, anisotropy, arbitrary structure profiles, and flexibilities of integration.

As shown in Fig. 2.7, the fabrication of MEMS metasurfaces has eight different processes. First, 0.5-μm Al thin film is deposited on the surface of a silicon on insulator (SOI) wafer, the size of which is chosen based on the standard of ion etching equipment. The photoresist (PR) is then spin-coated on the Al thin film. A photomask is used to define the meta-molecules pattern on the PR using ultraviolet illumination. After PR development, the Al patterns are realized by the ion etching process. This step can also be realized by using lift-off processes, which is not detailly discussed in this chapter. Then a hard mask is used to fabricate the MEMS comb-drive actuator and the supporting frame of the meta-molecules. The device layer of the SOI wafer is etched by the DRIE process, while the PR protects the microstructures patterned on the SiO_2 layer. The plasma-enhanced chemical vapor deposition (PECVD) dioxide is

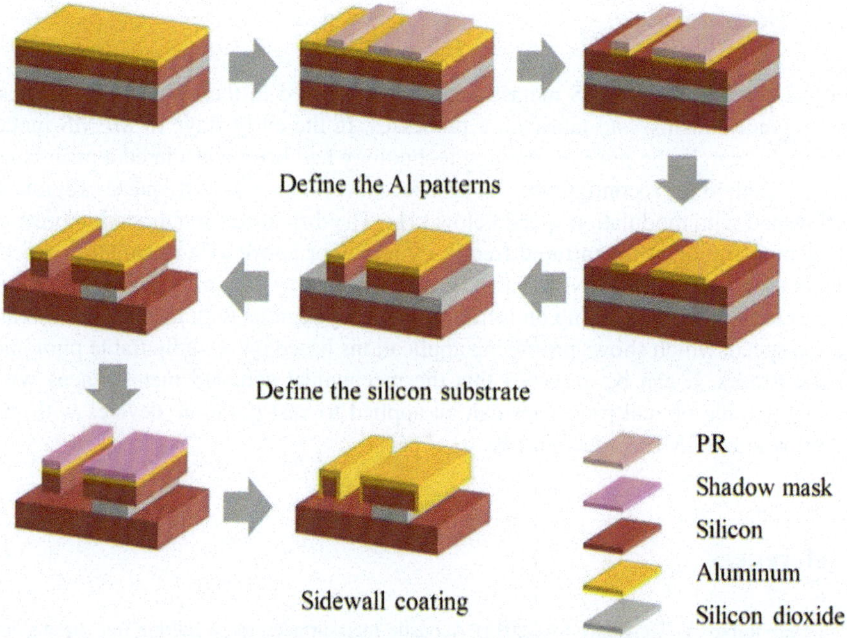

Fig. 2.7 The fabrication processes of a MEMS metasurface driven by electrostatic forces. The metasurface is working at the THz region, and the meta-molecules are composed of subwavelength patterns of the aluminum thin film. The ion etching processes are applied to achieve aluminum patterns on an SOI wafer. The supporting substrates of the meta-molecules are fabricated by using DRIE processes, while the sidewall coating is applied to realize the ohmic conducting of open and closed states for meta-molecules

applied to cover the surfaces and sidewalls of the etching trenches so that the notching effects are minimized for the sake of electromagnetic coupling between adjacent semi-squared rings. The supporting frames are released by using buffered oxide etchant (BOE) to free the movable parts of the meta-molecules with a well-controlled concentration and releasing time. The supporting frames of the fixed and movable parts of meta-molecules are under the same releasing process, which is precisely designed to release the MEMS actuator and the movable frame while the fixed frames can not. The reported release time of 3-μm supporting frames is approximately 15 min, which depends on the meta-molecules design, the thickness of the SOI working layer, the concentration of BOE, etc. It is the most critical process for the MEMS metasurface fabrication, which requires a lot of try and error works for the fabrication parameters. In the final step, the sidewall coating is applied for the ohmic conduction between the movable and fixed parts of the meta-molecules. A shadow mask protects the meta-molecules during the Al deposition along the side walls. The Al can bridge the gap between the meta-molecules deposited on the sidewall. This step is optional based on the design of meta-molecules.

2.6 Summary

In this chapter, the MEMS metasurfaces are discussed in their physics principles, tuning mechanisms, and fabrication processes. In the early stage of MEMS metasurface research, the micromachining technology had been considered a promising tuning method for reconfigurable metasurfaces working at THz frequency region or below and with modulation speed below kHz. The difficulties in fabrication prevent its demonstration at the infrared frequency region or above kHz modulation speed. Recent works on MEMS/NEMS metasurfaces have now demonstrated the reconfigurable metasurfaces working on infrared frequency regions with much faster modulation speed, which shows promising applications based on reconfigurable photonic metasurfaces. It can be expected that the mechanical actuated metasurfaces with reconfigurable optical properties can be applied to vast photonic devices with the development of MEMS technology.

References

1. Díaz-Rubio A, Tretyakov SA (2017) Acoustic metasurfaces for scattering-free anomalous reflection and refraction. Phys Rev B 96(12):125409
2. Tsitsas NL, Valagiannopoulos CA (2017) Anomalous reflection of visible light by all-dielectric gradient metasurfaces [Invited]. J Opt Soc Am B 34(7):D1–D8. https://doi.org/10.1364/JOSAB.34.0000D1
3. Xie Y, Yang C, Wang Y, Shen Y, Deng X, Zhou B, Cao J (2019) Anomalous refraction and reflection characteristics of bend V-shaped antenna metasurfaces. Sci Rep 9(1):6700. https://doi.org/10.1038/s41598-019-43138-1
4. Akselrod GM, Huang J, Hoang TB, Bowen PT, Su L, Smith DR, Mikkelsen MH (2015) Large-area metasurface perfect absorbers from visible to near-infrared. Adv Mater 27(48):8028–8034. https://doi.org/10.1002/adma.201503281
5. Alaee R, Albooyeh M, Rockstuhl C (2017) Theory of metasurface based perfect absorbers. J Phys D Appl Phys 50(50):503002. https://doi.org/10.1088/1361-6463/aa94a8
6. Li Y, Lin J, Guo H, Sun W, Xiao S, Zhou L (2020) A tunable metasurface with switchable functionalities: from perfect transparency to perfect absorption. Adv Opt Mater 8(6):1901548. https://doi.org/10.1002/adom.201901548
7. Song Q, Zhang W, Wu PC, Zhu W, Shen ZX, Chong PHJ, Liu A-Q (2017) Water-resonator-based metasurface: an ultrabroadband and near-unity absorption. Adv Opt Mater 5(8):1601103. https://doi.org/10.1002/adom.201301103
8. Yao Y, Shankar R, Kats MA, Song Y, Kong J, Loncar M, Capasso F (2014) Electrically tunable metasurface perfect absorbers for ultrathin mid-infrared optical modulators. Nano Lett 14(11):6526–6532. https://doi.org/10.1021/nl503104n
9. Gao H, Pu M, Li X, Ma X, Zhao Z, Guo Y, Luo X (2017) Super-resolution imaging with a Bessel lens realized by a geometric metasurface. Opt Express 25(12):13933–13943. https://doi.org/10.1364/OE.25.013933
10. Han F-Y, Huang T-J, Yin L-Z, Liu J-Y, Liu P-K (2018) Superfocusing plate of terahertz waves based on a gradient refractive index metasurface. J Appl Phys 124(20):204902. https://doi.org/10.1063/1.5050696
11. Liu T, Yang S, Jiang Z (2016) Electromagnetic exploration of far-field super-focusing nanostructured metasurfaces. Opt Express 24(15):16297–16308. https://doi.org/10.1364/OE.24.016297

12. Driscoll T, Andreev GO, Basov DN, Palit S, Cho SY, Jokerst NM, Smith DR (2007) Tuned permeability in terahertz split-ring resonators for devices and sensors. Appl Phys Lett 91(6):062511. https://doi.org/10.1063/1.2768300
13. Minovich AE, Miroshnichenko AE, Bykov AY, Murzina TV, Neshev DN, Kivshar YS (2015) Functional and nonlinear optical metasurfaces. Laser Photonics Rev 9(2):195–213. https://doi.org/10.1002/lpor.201400402
14. Nauman M, Yan J, de Ceglia D, Rahmani M, Zangeneh Kamali K, De Angelis C, Neshev DN (2021) Tunable unidirectional nonlinear emission from transition-metal-dichalcogenide metasurfaces. Nat Commun 12(1):5597. https://doi.org/10.1038/s41467-021-25717-x
15. Arbabi E, Arbabi A, Kamali SM, Horie Y, Faraji-Dana M, Faraon A (2018) MEMS-tunable dielectric metasurface lens. Nat Commun 9(1):812. https://doi.org/10.1038/s41467-018-03155-6
16. Manjappa M, Pitchappa P, Singh N, Wang N, Zheludev NI, Lee C, Singh R (2018) Reconfigurable MEMS Fano metasurfaces with multiple-input–output states for logic operations at terahertz frequencies. Nat Commun 9(1):4056
17. Meng C, Thrane Paul CV, Ding F, Gjessing J, Thomaschewski M, Wu C, Bozhevolnyi Sergey I. Dynamic piezoelectric MEMS-based optical metasurfaces. Sci Adv 7(26):eabg5639. https://doi.org/10.1126/sciadv.abg5639
18. Kwon H, Faraon A (2021) NEMS-tunable dielectric chiral metasurfaces. ACS Photonics 8(10):2980–2986. https://doi.org/10.1021/acsphotonics.1c00898
19. Lu C, Hu X, Shi K, Hu Q, Zhu R, Yang H, Gong Q (2015) An actively ultrafast tunable giant slow-light effect in ultrathin nonlinear metasurfaces. Light: Sci Appl 4(6):e302–e302. https://doi.org/10.1038/lsa.2015.75
20. Manjappa M, Chiam S-Y, Cong L, Bettiol AA, Zhang W, Singh R (2015) Tailoring the slow light behavior in terahertz metasurfaces. Appl Phys Lett 106(18):181101. https://doi.org/10.1063/1.4919531
21. Papaioannou M, Plum E, Valente J, Rogers ETF, Zheludev NI (2016) Two-dimensional control of light with light on metasurfaces. Light: Sci Appl 5(4):e16070–e16070. https://doi.org/10.1038/lsa.2016.70
22. Gao W, Ota H, Kiriya D, Takei K, Javey A (2019) Flexible electronics toward wearable sensing. Acc Chem Res 52(3):523–533
23. Gates Byron D (2009) Flexible electronics. Science 323(5921):1566–1567. https://doi.org/10.1126/science.1171230

Chapter 3
Microfluidic Metasurfaces

Metamaterials and metasurfaces are artificial materials consisting of meta-molecules that are rationally designed to mimic the atoms and molecules in natural materials. Metasurfaces are two-dimensional (2D) materials with all the meta-molecules spatially arranged on a surface, which is typically subwavelength in-depth regarding the propagation direction of the incident electromagnetic waves. Recently, metasurfaces have been intensively studied due to their ability to control the amplitude, phase, polarization states, and wavefront of the incident light, leading to vast applications, including optical switches, flat lenses, beam steering, etc. The flexibilities of the meta-molecule geometry design offer a new degree of freedom to design extraordinary electromagnetic properties of artificial materials. More importantly, the material properties can be tuned, reconfigured, or even redefined by changing the geometries of meta-molecules which not only offers a new approach to control the material properties in terms of optical constants, i.e., permeabilities and permittivities, but also paves the way to design tunable and reconfigurable photonic devices based on the spatial reconfiguration and geometric tuning of meta-molecules.

On the other hand, tunable and flexible photonic devices have now been demonstrated in vast applications based on multidisciplinary research on biology and photonics. For example, photonic devices with flexible substrates can be applied to artificial surfaces and tissues, which have promising applications on bio-sensors, artificial skins, stealth coating, wearable health care devices, etc. Electronic devices based on flexible substrates have already been studied for decades, which paves the way for the fabrication techniques of flexible photonic devices. However, the flexible photonic and electronic devices are still longing for advanced materials with designable electromagnetic/mechanical/chemical properties, which rely on artificial material research covering many topics of fundamental interest. Although the development of integrated photonics and nanomaterials have been studied for years, flexible photonic devices are limited by the choices of natural materials with desired optical and mechanical properties. For example, it is challenging to find a transparent material at broadband working frequencies, e.g., the infrared region, which is stretchable and compatible with the matured fabrication techniques.

W. Zhu and A.-Q. Liu, *Metasurfaces: Towards Tunable and Reconfigurable Meta-devices*, Microfluidics and Nanophotonics: Science and Engineering 1,
https://doi.org/10.1007/978-981-19-6925-6_3

The flexible and tunable metasurfaces are recently proposed based on similar fabrication techniques as their counterparts of flexible electronics. For example, pattern transfer techniques [1–4] are applied to transfer the solid meta-molecules made of metal or high-refractive-index dielectric materials to stretchable substrates made of soft materials. Meta-molecules made of solid-state materials are patterned on soft-material substrates forming half-rigid half-flexible artificial materials for tunable and reconfigurable material properties, which are often called flexible metasurfaces. Another widely demonstrated approach is liquid crystal metasurfaces which are realized by changing the refractive indexes of surrounding media. The optical properties of the meta-molecules can be tuned by changing the surrounding media, e.g., liquid crystal. The liquid crystal metasurfaces are typically made of multi-layered structures, including a rigid substrate, a meta-molecule layer surrounded by liquid crystal, and the containing/control structures for liquid crystals. Most liquid crystal metasurfaces have tunable optical properties but without any stretchabilities.

Two different approaches are demonstrated for metasurfaces with both tunable optical properties and mechanical stretchabilities. The first is to use droplet array as meta-molecules anchored on flexible substrates. The droplet-based metasurfaces are typically demonstrated at the radio frequency (RF) region since the high-refractive-index liquid are rare in other working frequencies. The other approach regulates liquid materials, i.e., water, galinstan, and mercury, with microfluidic systems and is called microfluidic metasurfaces. The metasurfaces with designable and reconfigurable optical/mechanical properties are promising candidates for wearable devices and tunable flat optics. The electromagnetic properties can be easily controlled by changing the geometries and the periods of meta-molecules. It can be expected that flexible metasurfaces can be applied to vast applications, including tunable flat lens, dynamic beam steering, stealth coating, electronic wallpaper, and display, just to name a few. This chapter discusses metasurfaces based on soft materials and microfluidic control systems.

3.1 Metasurfaces Based on Soft Materials

A Brief Introduction to Dielectric Metasurfaces

The optical properties of metal-based mete-molecules typically result from the induced electronic currents with electromagnetic incident waves, as shown in Fig. 3.1a. For example, surface plasmonic polariton (SPP) resonances can be induced at metal-dielectric interfaces consisting of meta-molecules with proper momentum conservation conditions. On the other hand, dielectric meta-molecules are also proposed and demonstrated, which control the phase and amplitude of the incidence by optical resonances induced by the electrical displacements of atoms, as shown in Fig. 3.1b. A widely demonstrated example is the dielectric meta-molecules based on Mie resonances. The displacement currents can induce multipole/monopole resonances for both electric and magnetic fields. As a result, both permittivities and

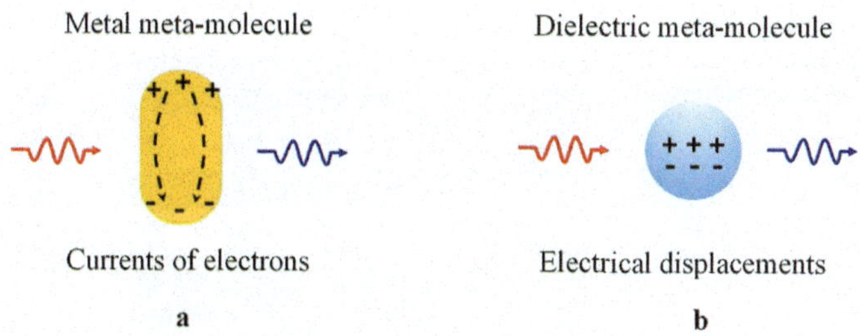

Fig. 3.1 Comparison between **a** metal and **b** dielectric meta-molecules

permeabilities of the dielectric metasurfaces can be designed by changing the geometries of the dielectric meta-molecules. In this section, dielectric spherical particles are discussed as an example of meta-molecules based on Mie resonance.

The exact Mie solution of the diffraction problem [5, 6] is often used to explain the optical properties of meta-molecules with a diameter less than a quarter of the incident wavelength. The scattered field of an isolated dielectric sphere can be decomposed into multipole electric and magnetic resonances where the electrical and magnetic fields can be written as a_m and b_m, respectively.

$$a_m = \frac{n\psi_m(nx)\psi'_m(x) - \psi_m(x)\psi'_m(nx)}{n\psi_m(nx)\xi'_m(x) - \xi_m(x)\psi'_m(nx)} \tag{3.1}$$

$$a_m = \frac{n\psi_m(nx)\psi'_m(x) - \psi_m(x)\psi'_m(nx)}{n\psi_m(nx)\xi'_m(x) - \xi_m(x)\psi'_m(nx)} \tag{3.2}$$

where n is the refractive index of the dielectric sphere and $x = k_0 r_0$, k_0 is the wavenumber of incident electromagnetic wave, r_0 is the radius of the dielectric sphere, $\psi_m(x)$ and $\xi_m(x)$ are the Riccati-Bessel functions. a_m and b_m are the scattering coefficient related to the electric and magnetic responses of the dielectric sphere, respectively.

Similar to the discussion in Sect. 2.1, the optical properties of metasurfaces depend on both the meta-molecules and their mutual couplings. The effective optical properties of materials consisting of an array of dielectric spheres embedded in a homogeneous medium were given by Lewin back in 1946 [7]. The effective permittivity and permeability can be derived by combining the Mie scattering with Clausius–Mossotti's effective medium theory.

$$\varepsilon_{eff} = \varepsilon_1\left(1 + \frac{3\nu_f}{\frac{F(\theta)+2b_e}{F(\theta)-b_e} - \nu_f}\right) \tag{3.3}$$

$$\mu_{eff} = \mu_1\left(1 + \frac{3\nu_f}{\frac{F(\theta)+2b_m}{F(\theta)-b_m} - \nu_f}\right) \tag{3.4}$$

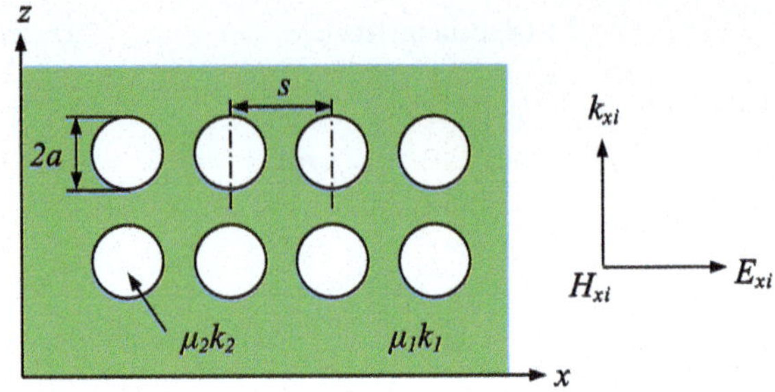

Fig. 3.2 Schematic of a dielectric sphere array embedded in a homogenous medium

where $F(\theta) = \frac{2(\sin\theta - \theta\cos\theta)}{(\theta^2-1)\sin\theta + \theta\cos\theta}$ is a resonant function, $b_e = \frac{\varepsilon_1}{\varepsilon_2}$, $b_m = \frac{\mu_1}{\mu_2}$, the filling fraction of the spherical dielectric particles $\nu_f = \frac{4}{3}\pi\left(\frac{r_0}{s}\right)^3$, $\theta = k_0 r_0 \sqrt{\varepsilon_2 \mu_2}$, and s is the lattice constant as shown in Fig. 3.2.

It should be noted that the constitutive parameters retrieved from Lewin's model were formulated only considering the dielectric particles resonating either in the first or second resonance modes of the Mie series. Lewin's model only considers a particular resonance mode of Mie resonances, which is inaccurate with multiple resonances with nontrivial effects. However, it shows a simplified method for calculating dielectric metasurfaces' effective optical parameters. The first demonstration of dielectric metamaterials can be traced to 2003. Then dielectric meta-molecules have been widely demonstrated in metamaterials and metasurfaces.

Compared with metal meta-molecules, dielectric meta-molecules have lower instinct loss which can be applied to metasurfaces to achieve high diffraction efficiencies. For example, silicon meta-molecules are often used for metasurfaces working in the infrared region, which shows vast applications on blazed gratings, beam steering devices, metalens, etc. [8–14] Although the absorption of silicon in the mid-infrared region is relatively higher than the germanium and ZeSn, silicon-based metasurfaces can still achieve higher than 80% diffraction efficiency due to the subwavelength thickness. The diffraction efficiency can even reach near unity, which was demonstrated using dielectric Huygens's metasurfaces. The near-unity transmission and reflection of homogeneous metasurfaces are also demonstrated by achieving the optical impedance-matching conditions, which are widely applied to perfect absorbers. However, the impedance-matching condition requires the control of magnetic and electric resonances simultaneously at a working wavelength which is quite difficult to be applied to broadband devices based on metasurfaces. On the other hand, inhomogeneous metasurfaces are often proposed to tailor the incident wavefront with meta-molecules designed to cover a 2π range of abrupt phase change. The Mie-type dielectric meta-molecules can have magnetic and electric resonances at the same frequency region, which is essential to design larger-than-π phase change and optical impedance matching.

Other materials are used to design dielectric metasurfaces when the working frequency goes up to the visible range. For example, gallium phosphide (GaP), silicon mononitride (SiN), and titanium dioxide (TiO_2) have much lower intrinsic absorption at the visible range compared with silicon. The fabrication technologies are also needed to be considered for the choices of materials. Many TiO_2 metasurfaces have now been demonstrated to work at the visible frequency region. For example, a high numerical aperture (NA) flat lens is demonstrated using TiO_2 metasurface, which shows a NA of 0.8 and total efficiency of 86%. It should be pointed out that metal meta-molecules are often used in metasurfaces working at radiofrequency or below since the metal functions as a perfect electric conductor (PEC) at the lower frequency region.

Liquid Crystal Metasurfaces

Liquid crystals are often applied to tunable and reconfigurable optical devices since their refractive indexes can be tuned by thermal/optical/electrical effects. The optical properties of liquid crystals can be controlled by the arrangement of their molecules, which show high anisotropy, birefringence, and nonlinearity. For example, the permittivity of Lyotropic liquid crystals can be changed by both temperature variation and their concentration in the solvent. Nematic liquid crystals can be controlled by temperature variation, external pressure, and electromagnetic fields, which are widely used in display devices.

Liquid crystal metasurfaces are multilayered structures with liquid crystal as surrounding media, which can be used to tune the resonance frequencies of meta-atoms[15–17]. Many applications have now been demonstrated using liquid crystal metasurfaces, such as tunable filters, optical switches, beam steering devices, etc. Pioneer works on liquid crystal metasurfaces are demonstrated in the GHz region. The metasurfaces are tuned by DC voltage ranging from 0 to 100 V with a sub-second response time due to the large sizes of the devices. Many applications have now been demonstrated using liquid crystal metasurfaces. Both magnetic and electric dipole resonant frequencies are shifted corresponding to the orientation of liquid crystal molecules, which functions as a tunable filter. The optical switch can be realized by switching the orientation of liquid crystal molecules between two orthogonal directions. Liquid crystals are injected into a multi-layered metasurface. The absorption band of the metasurface is tuned by changing the coupling between the meta-atom layer and the ground plane. Recently, a THz tunable absorber was demonstrated based on a liquid–crystal metasurface with porous graphene as electrodes to control the refractive index of liquid crystals. The absorption frequency is tuned from 0.75 THz to 1 THz with a larger-than-80% absorption efficiency. Tunable lens functions are also demonstrated using liquid crystal metasurfaces with all the meta-molecules tuned by the surrounding media.

The tuning speed of the liquid crystal metasurfaces is typically less than kHz, which is capped by the rotation time of the liquid crystal molecules. Therefore, it is challenging to be applied to fast tuning devices such as optical switches. Also, the anisotropic properties of the liquid crystals induced by the tuning process can affect metasurfaces' optical response, which increases the design difficulties of polarization-independent devices. However, meta-molecules composed of liquid crystals may be applied to many extraordinary functionalities with the development of microfluidic technologies.

Flexible Substrate Metasurfaces

The technology of flexible electronics has been developed for decades, which paves the way for system-level integration of flexible metasurfaces. Recently, a system-on-plastic device based on flexible electronics has been demonstrated with three different layers of components, proving that the integration of solid-state electronic components and the soft substrate is now possible. However, other technical issues need to be considered for devices based on flexible metasurfaces besides integration difficulties. The tunable and reconfigurable metasurfaces based on flexible substrates are tuned by the variation of the geometries and lattice constants of the meta-molecules, which rely on the stretchability of the substrates. For example, a flexible metasurface is composed of gold meta-molecules anchored on a Polydimethylsiloxane (PDMS) substrate, which has surface plasmonic polariton (SPP) enhanced absorption at the infrared region [18]. The stretching of the substrate results in the tuning of meta-molecules and the lattice constant simultaneously. The central wavelength of the absorption band is demonstrated to have a tunability of 400 nm since the SPP resonance frequency is highly dependent on both the meta-molecules geometry and the lattice constant.

The flexible substrates also offer a new approach to designing artificial materials with nonlinear optical properties. One demonstration is a magneto-elastic metasurface with a nonlinear optical response arising from the deformation of the substrate between two layers of SRR meta-molecules [19]. The substrate is deformed by Ampere forces induced by the electronic currents on the SRR meta-molecules, which can be either attractive or repulsive depending on the resonance modes of the electrons. The magneto-elastic is demonstrated in the radiofrequency region, which is followed by many works on metasurfaces actuated by optical forces. It can be expected that this technology will be applied to both artificial optical materials with large nonlinearity and nanomotors driven by photon energy.

Portable and wearable photonic devices are another exciting research topic based on flexible metasurfaces, which enable the core technology of compact and flexible optical components. Flexible metasurfaces have twofold advantages in terms of practical applications. The first one is the large tunability due to the mechanical actuation. The other is that the soft-material substrate can be designed to accommodate arbitrary surfaces, which is essential for wearable devices. Although the stability and repeatability are not satisfactory for most particle applications, vast applications on portable and wearable photonic devices have been demonstrated using flexible metasurfaces, including flexible display, tunable lens, wearable sensor, etc.

Wavefront reconfiguration is also demonstrated using flexible metasurfaces. One example is a tunable flat lens consisting of a gold nano-rod array and a PDMS substrate. The flat lens works on the visible frequency region with the focal length tuned from 150 to 250 μm when the PDMS substrate is stretched. The meta-molecules are designed to maintain the hyperbolic phase profile during the tuning of the substrate. Although the flexible metasurfaces are compact, the mechanical actuation system is bulky and difficult to integrate into chip-scale devices. MEMS actuators are recently applied to tune the flexible metasurface in a well-controlled manner. Four electrostatic MEMS actuators are used to stretch the flexible substrate while maintaining the rotational symmetry of the lattice for flat lens functionality. Combining MEMS technologies and flexible metasurfaces is one big step towards practical applications on compact devices.

The fabrication technologies for flexible substrate electronic devices with nanoscale patterns have been developed for decades. The combination of solid-state nanopatterns and soft-material substrates requires fabrication technologies with high sophistication developed by multidisciplinary research. The fabrication processes based on soft lithography are shown in Fig. 3.3, widely applied to flexible metasurfaces. First, a polymer layer is spin-coated on a solid-state silicon substrate called a carrier substrate. Then chromium layer is deposited on the polymer as an adhesive layer between the meta-molecules and the polymer layer. The metal layer is deposited on the adhesive layer, and the photoresist (PR) layer is spin-coated on top of it. The meta-molecules patterns are defined using photolithography processes. Finally, the flexible metasurface consists of a polymer layer, and a meta-molecules array is ripped off from the carrier substrate. Besides photolithography, the meta-molecules can be fabricated using other technologies, including electron beam lithography (EBL), focus ion beam (FIB), electroplating, laser microlens array lithography, direct laser writing (DLW), etc.

The replica molding technology [20, 21] is also applied to the fabrication of flexible metasurfaces, which has great potential for the massive production of nanoscale patterns with soft-material substrates. First, a solid-state stamp structure is fabricated using the technologies mentioned above. Then the polymer layer is spin-coated on a carrier substrate. The stamp structure is pressed on the polymer layer to transfer the meta-molecule structures to the soft-material substrate, which is removed after the post-curing process. This technology is suitable for dielectric metasurfaces with the meta-molecules consisting of photoresist or other polymer materials.

Another widely used fabrication technology is transfer printing for solid-state meta-molecules on the soft-material substrate. First, the meta-molecules are fabricated on a solid-state substrate. Then a picked-up process is applied to transfer the meta-molecules to the soft-material substrate using an elastomeric stamp. The adhesive layers of the solid-state substrate and the soft-material substrate must be

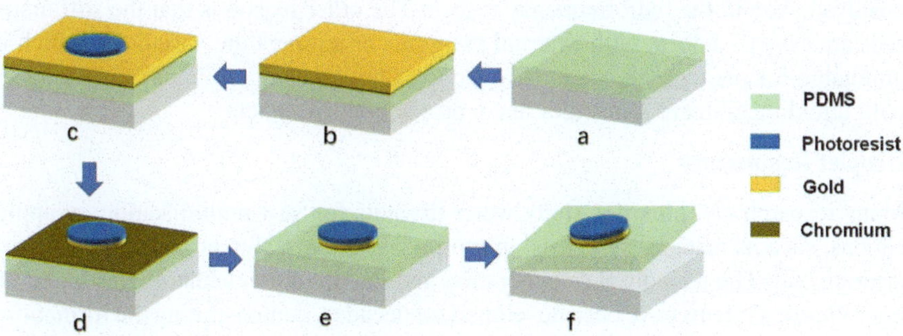

Fig. 3.3 Fabrication processes of flexible metasurfaces based on soft lithography

appropriately chosen for the transfer processes, typically realized by thermal treatments. Complicated designs can be achieved by combing several different fabrication technologies.

Like their counterparts in flexible electronics, flexible metasurfaces have been widely demonstrated using solid-state meta-molecules and soft substrates. The flexible metasurfaces can be tuned by stretching the soft substrates, resulting in the variation of the geometry and spacing of meta-molecules. Therefore, the precise control of the flexible metasurfaces becomes very difficult since the electromagnetic responses depend on the meta-molecules and their mutual coupling. Although the metasurfaces are small in size, the control systems of flexible metasurfaces are typically bulky and difficult to apply to compact devices. More importantly, the meta-molecules are composed of solid-state materials, such as metal or dielectric materials with a high refractive index, which is likely to be cracked during the stretching of the substrate. Therefore, many technical issues have to be solved before the flexible metasurfaces can be applied to practical applications. It can be expected that those technical issues will be solved with the development of flexible electronics and photonics, which are driven by the desired technologies for wearable and reconfigurable devices from the market.

3.2 Microfluidic Metasurfaces

Although flexible metasurfaces offer large tuning ranges on optical properties based on mechanical actuation, soft materials are not the first choice for tunable and reconfigurable optics due to the following two reasons. First, the electron currents or SPP effect often require metal structures with high conductivities, while the Mie resonances require a high refractive index. Soft and nonabsorptive materials typically have low refractive indexes or conductivities, which, therefore, are not proper choices for meta-molecules design. Soft materials, e.g., water, are found to have a high permittivity in the radiofrequency region, which has now been applied to RF metasurfaces. However, it is challenging to find soft materials with a high refractive index (> 3) in other frequency regions. Besides, the absorption problems remain unsolved even in the radiofrequency region. The other reason is that the soft materials are easy to deform with external pressures or temperature variations, which is unreliable for photonic devices. This section introduces microfluidic metasurfaces with liquid metal meta-molecules and a proper control system.

Droplet Resonators

Many research communities study water droplets for vast multidisciplinary applications, such as material synthesis, microfluidic systems, ship engine design, inkjet printing, etc. The nanodroplets generation and manipulation technologies based on microfluidic systems have been developed for decades, demonstrating the formations of periodical droplet arrays using water, oil, liquid crystal, etc.

Recently, droplet formation technologies have been applied to metasurfaces working on the radiofrequency region. For example, a radiofrequency resonator is designed using metal cavities filled with water, which is an elliptical cylinder tuned by the filling factor of water. The rotation of the metasurfaces can tune the transmission coefficient of the homogeneous metasurface. Another droplet-based metasurface is demonstrated as a perfect absorber, which is realized by droplet formation on a substrate with hydrophilic and hydrophobic treatments. The periodical arranged hydrophilic and hydrophobic regions of the substrate result in the formation of a droplet array when water is poured on it. This technology can be applied to both solid-state and soft-material substrates, including paper, polyethylene terephthalate (PET), Flame Retardant 4 (FR-4), glass, etc. The shapes of the water droplets, i.e., the sizes and contact angles, can be predesigned by the treatment conditions, resulting in absorption bands' design. However, the free-stand droplets are very difficult to meet the requirements on the reliability of the device level and cannot be dynamically tuned.

Metasurfaces made of soft materials only, i.e., PDMS and water, are also demonstrated using microfluidic systems [22–24]. The PDMS substrate is patterned with a micro-reservoirs array, which is used to control water droplets' shapes and spatial distributions. This metasurface is demonstrated as a perfect absorber working in the radiofrequency region. Near-unity absorptivity is measured at the central frequency due to the well-defined meta-molecules based on the microfluidic droplets array. The combination of microdroplet technologies and metasurfaces paves the way for vast applications in wearable and tunable devices.

Microfluidic Control Systems and Tuning Mechanisms

In the early 1980s, microfluidic technologies are applied to inkjet printheads, DNA chips, lab-on-a-chip devices, etc. The 40-years research on microfluidic technologies enables the generation and precise manipulation of droplets of nanoliter or less, which is an ideal control system for meta-molecules based on droplets. In 2002, Prof. Quake demonstrated a microfluidic control system for large-scale integration of pneumatic valves, as shown in Fig. 3.4. In 2015, the microfluidic control system based on a pneumatic valve multiplexer was applied to tunable and reconfigurable metasurfaces, which are named microfluidic metasurfaces.

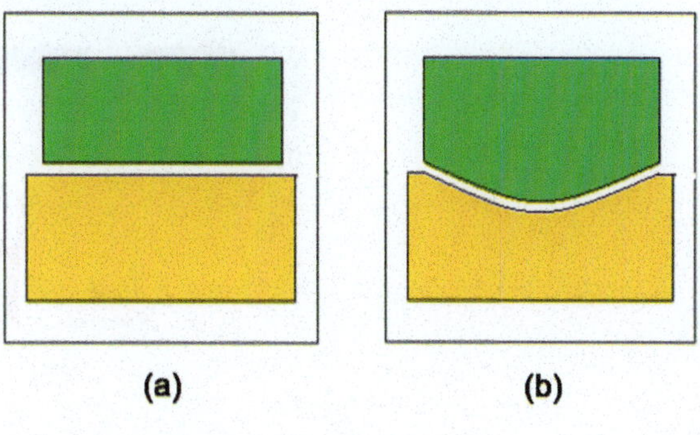

Fig. 3.4 The working principle of pneumatic valves. **a** On state and **b** off state

The droplets within microfluidic channels are often controlled by the air pressure from external pumps, which is a noncontact tuning method. The pneumatic valves based on microfluidic channels are designed to regulate the control pressure so that the droplets with the microfluidic channels can be tuned individually by selectively sending the pressure to the target droplet.

Here, the pneumatic valves consist of two layers of PDMS channels, which are the control and structure layers, respectively. As shown in Fig. 3.4, the control layer (green) is composed of micro-valves on the top of the micro-channels (yellow), while the white parts are the PDMS membranes. The PDMS membrane between the control and structure layer is planar when the air pressure within the control valves and structure channels are equal. Otherwise, the unbalanced pressure between each side will deform the PDMS membrane. As a result, the deformation of the PDMS membranes of the structure layer will squeeze the air into the reservoirs with droplets inside and change the geometric shapes of the droplets.

The pneumatic valve multiplexer is applied to the droplets array to realize the random access metasurface, the meta-molecules of which can be tuned individually. First, the random access metasurface is demonstrated as a tunable flat lens using a ternary multiplexer system, as shown in Fig. 3.5.

Figure 3.5 shows a simplified ternary multiplexer control system with 2 bits to control nine structure channels (red) individually using six control channels (blue and green). The control layers are composed of pneumatic valves linked by microfluidic channels. The pneumatic valves are micro reservoirs with a larger dimension in width than the microfluidic channels. Once pumped with external pressures, larger forces will be applied to the structure channels where the pneumatic valves are on the top (marked with a cross). As a result, the microfluidic channels within the structure layer will be blocked by the PDMS membrane deformation induced by the pneumatic valves. Although the blocking channel comes across other microfluidic channels in

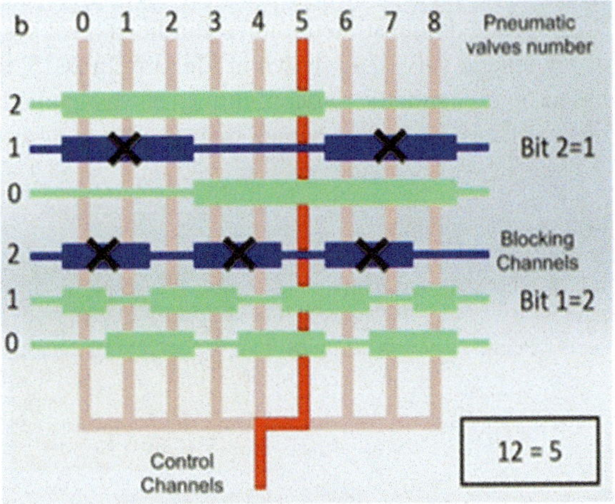

Fig. 3.5 Schematic of a simplified ternary multiplexer for a microfluidic metasurface control system

the structure layer, the PDMS membrane deformation area is not large enough to block the whole microfluidic channel, which remains in the open state.

Soft lithography technologies are widely applied to the fabrication of microfluidic metasurfaces, which are matured processes for patterning micro/nano structures on the soft substrate. Figure 3.6 shows the fabrication processes of a tunable flat lens based on a microfluidic metasurface [25]. Here the SU8 is used as the master of the PDMS structures. The SU8 photoresist is spin-coated on a silicon substrate, as shown in Fig. 3.6a and Fig. 3.6b. Then the SU8 master structures are defined by the UV exposure with a mask. The PDMS microstructures are realized by pouring the liquid PDMS on the SU8 master structures. The liquid PDMS becomes solid after approximately 2 h of baking at 120 °C or 24 h at room temperature. Here, the height of the microfluidic channel equals the thickness of the SU8 layer, which can be controlled by the choices of the SU8 photoresist and the spin-coat parameters. The PDMS layer's thickness can also be defined by controlling the height of liquid PDMS on the master structures. The spin coater can also be used for the submillimeter height of the PDMS layer. The PMMA layer is used as the supporting substrate for the microfluidic metasurface, which is spin-coated with a PDMS layer for better adhesion. The tunable and reconfigurable devices with large dimensions can be achieved by forming a microfluidic metasurfaces array, as shown in Fig. 3.6j. Microfluidic metasurfaces have now become an exciting research topic for tunable and reconfigurable devices working in GHz and THz regions. The microfluidic technologies extend the choices of meta-molecules compositing materials from solid-state to liquid-state. More importantly, droplet generation and manipulation capabilities offer great flexibility for the meta-molecules reconfiguration, which enables vast applications.

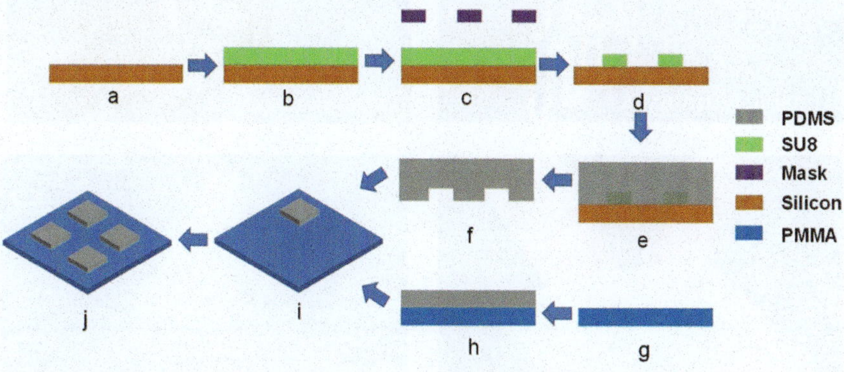

Fig. 3.6 Fabrication processes of a tunable flat lens based on microfluidic metasurface

3.3 Applications of Microfluidic Metasurfaces

The individually tuning of meta-molecules by microfluidic multiplexer enables a new category of metasurfaces based on soft materials, which are applied to tunable and reconfigurable meta-devices and function switchable meta-devices. This section gives a brief introduction to microfluidic metasurfaces, which includes reconfigurable optical properties, the amplitude and polarization state control of the incidence, dynamic wavefront reconfiguration, etc.

Figure 3.7a shows the schematic of a multilayered microfluidic metasurface with tunable optical chirality in the THz frequency region, which is realized by controlling the space-parity symmetry [26]. The microfluidic metasurface is composed of soft materials only, which can be tuned by either changing the refractive index of the liquid within the micro-channels or simply stretching the metasurface. The metasurface is a thin film with a 500-μm thickness, which is 1.5 times larger than the working wavelength. The incident THz wave with circular polarization is measured to have a polarization rotation angle from 0° to 16.9° by changing the refractive index of the liquid within the micro-channels, which shows an optical chirality tuning from a non-chiral state to a strong-chiral state.

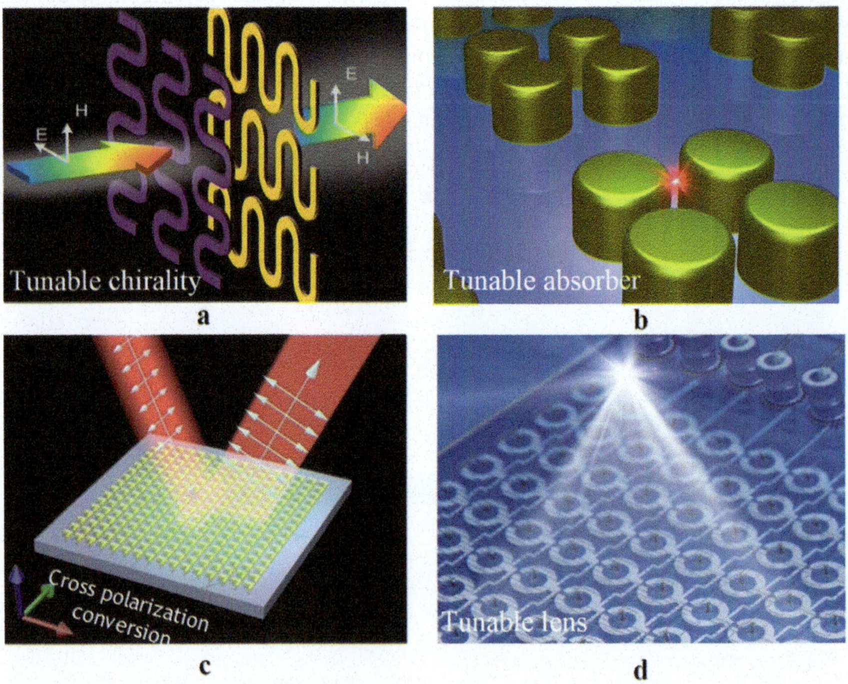

Fig. 3.7 Schematics of tunable and reconfigurable devices based on microfluidic metasurfaces. **a** Microfluidic metasurfaces with tunable chirality in the THz region. **b** Tunable absorber with liquid metal droplets. **c** Reconfigurable polarization converter. **d** Tunable flat lens

Another demonstration of the THz broadband absorber with a tunable absorption band is shown in Fig. 3.7b, which is realized by a liquid metal droplets array with a microfluidic system. The meta-molecules are composed of four identical liquid metal pillars with strong mutual couplings and anchored on a ground plane of liquid metal. The absorption is achieved by the destructive interference of the reflected THz waves controlled by the height of the meta-molecules. In the experiment, the height of the pillars is controlled by the external pressure applied to the metasurface where the complex microfluidic control system, e.g., the multiplex, is not a necessity. The central absorption frequency is tuned from 0.245 to 0.415 THz with an above 90% absorption efficiency, covering a full optical octave.

The microfluidic multiplexer control system enables individual geometry reconfiguration of meta-molecules in a large array, which is applied to random access microfluidic metasurfaces with dynamic wavefront control and switchable functionalities. A broadband multifunctional polarization converter is demonstrated using L-shaped liquid metal droplets, as shown in Fig. 3.7c. The microfluidic metasurface is switched between different functionalities by changing the geometric shapes of the meta-molecules. The incident electromagnetic waves can be converted to different polarization states, i.e., linear, circular, and elliptic polarization states, which was demonstrated in the experiment. Additionally, the broadband and optical attenuation function is also demonstrated, which has potential applications on smart radar, optical attenuator, and light manipulation for quantum optics, etc.

Figure 3.7d shows random access reconfigurable metasurface for a tunable flat lens, which is realized by SRR liquid metal meta-molecules controlled by the microfluidic multiplexer. The meta-molecules are formed by microfluidic ring-shaped reservoirs partially filled with mercury, which are regulated using pneumatic valves. The microfluidic metasurface was demonstrated as a tunable flat lens with a focal length tuned from 10 to 30 cm, which works in the Ku band of the radiofrequency region.

3.4 Summary

At the very beginning, tunable metamaterials with active and switchable optical properties are proposed to compensate for the drawbacks of the metamaterials, such as high absorption loss and narrow working bandwidth, which are typically due to the strong resonance nature of the meta-molecules. Now, exciting and technologically important capabilities of tunable metamaterials range from tunable optical charity and controllable magnetic resonance to dynamic quantum effects, with applications across science and engineering from active photonic devices to biological sensors and imaging systems. Most tunable metamaterials research targeting nanophotonic circuits is limited by the availability of fast and highly responsive nonlinear media that react to the control signals by phase-changing or refractive index variation. It is challenging to deliver in nanoscale devices using electronic or molecular nonlinearities. The tuning range and speed are often limited by the saturation effects of the chosen nonlinear media and the sub-wavelength scale optical paths. On the other hand, structural reconfigurable metamaterials were demonstrated with a low switching speed of several Hertz and large meta-molecules sizes of millimeter-scale several years ago, which has meta-molecules now been minimized to submicron scale with tuning speed up to megahertz. The structural reconfigurable metasurfaces based on micromachined actuators or other tuning methods can be possible alternatives to active metamaterials with the help of the quick and widespread proliferation of new nanofabrication techniques. However, the capabilities of structural reconfiguration are always limited by the solid-based metal patterns, which cannot be changed once fabricated.

It can be anticipated that the fundamental breakthroughs in tunable metamaterials research will occur when combining the concept of metasurfaces with microfluidics technology. The pure liquid-based functional materials, which are named microfluidic metasurfaces, are feasible to manipulate light with engineered sub-wavelength structures based on fluidics, which has much more freedom in structural reconfiguration than those of tunable metamaterials and metasurfaces based on solid metal patterns. Furthermore, optofluidic technology offers a platform for controlling the fabric of "electromagnetic space" where the diffusion of liquid realizes the gradient refractive indices. Therefore, transformation optics based on microfluidic metasurfaces offers additional technological opportunities, which are not feasible with conventional homogeneous optical materials.

In the following chapters, recent advances in tunable metamaterials will be discussed, including structural reconfigurable metamaterials based on MEMS, flexible substrate metamaterials, and soft-material metasurfaces based on microfluidic control systems. For comparison, a brief introduction will be given to the active metamaterials based on nonlinear materials, sensitive to photon illumination, external electrical/magnetic fields, or temperature variations. The standard treatments of such problems will be discussed, along with their limitations, and the alternative tunable metamaterials based on MEMS actuation and flexible substrate will be delineated. A prospective study on microfluidic metasurfaces is presented, which is based on current microfluidic technologies widely used in optofluidic systems, highlighting the most promising approaches and the further work that remains to be done. Exciting developments in microfluidic metasurfaces in transformation optics are also discussed. An outlook of the important challenges that remain to be addressed will be given, such as the control systems might be applied to pure liquid-based meta-molecules, the limitation and possible improvement on the meta-molecule size and tuning speed, as well as intriguing new directions for the field, including single and random accessible tuning out of a massive meta-molecules array without downgrading the metasurfaces devices performances using electric circuits.

References

1. Xia Y, Zhao XM, Whitesides GM (1996) Pattern transfer: self-assembled monolayers as ultrathin resists. Microelectron Eng 32(1):255–268. https://doi.org/10.1016/0167-9317(95)00174-3
2. Odom TW, Love JC, Wolfe DB, Paul KE, Whitesides GM (2002) Improved pattern transfer in soft lithography using composite stamps. Langmuir 18(13):5314–5320. https://doi.org/10.1021/la0201691
3. Linghu C, Zhang S, Wang C, Song J (2018) Transfer printing techniques for flexible and stretchable inorganic electronics. NPJ Flexible Electron 2(1):26. https://doi.org/10.1038/s41528-018-0037-x
4. Walia S, Shah CM, Gutruf P, Nili H, Chowdhury DR, Withayachumnankul W, Sriram S (2015) Flexible metasurfaces and metamaterials: a review of materials and fabrication processes at micro- and nano-scales. Appl Phys Rev 2(1):011303. https://doi.org/10.1063/1.4913751
5. Tribelsky MI, Flach S, Miroshnichenko AE, Gorbach AV, Kivshar YS (2008) Light scattering by a finite obstacle and fano resonances. Phys Rev Lett 100(4):043903. https://doi.org/10.1103/PhysRevLett.100.043903
6. Brown RGW (1984) Absorption and scattering of light by small particles. Opt Acta 31:3. https://doi.org/10.1080/716099663
7. Lewin L (1947) The electrical constants of a material loaded with spherical particles. J Inst Electr Eng—Part III: Radio Commun Eng 94(27):65–68. https://doi.org/10.1049/ji-3-2.1947.0013
8. Staude I, Schilling J (2017) Metamaterial-inspired silicon nanophotonics. Nat Photonics 11(5):274–284. https://doi.org/10.1038/nphoton.2017.39
9. Kamali SM, Arbabi E, Arbabi A, Faraon A (2018) A review of dielectric optical metasurfaces for wavefront control. Nanophotonics 7(6):1041–1068. https://doi.org/10.1515/nanoph-2017-0129
10. Lin D, Fan P, Hasman E, Brongersma Mark L (2014) Dielectric gradient metasurface optical elements. Science 345(6194):298–302. https://doi.org/10.1126/science.1253213
11. Chen WT, Zhu AY, Capasso F (2020) Flat optics with dispersion-engineered metasurfaces. Nat Rev Mater 5(8):604–620. https://doi.org/10.1038/s41578-020-0203-3
12. Shalaginov MY, An S, Zhang Y, Yang F, Su P, Liberman V, Gu T (2021) Reconfigurable all-dielectric metalens with diffraction-limited performance. Nat Commun 12(1):1225. https://doi.org/10.1038/s41467-021-21440-9
13. Wu PC, Pala RA, Kafaie Shirmanesh G, Cheng W-H, Sokhoyan R, Grajower M, Atwater HA (2019) Dynamic beam steering with all-dielectric electro-optic III–V multiple-quantum-well metasurfaces. Nat Commun 10(1):3654. https://doi.org/10.1038/s41467-019-11598-8
14. Chen R, Shao Y, Zhou Y, Dang Y, Dong H, Zhang S, Ma Y (2022) A semisolid micromechanical beam steering system based on micrometa-lens arrays. Nano Lett. https://doi.org/10.1021/acs.nanolett.1c04493
15. Ma Z, Meng X, Liu X, Si G, Liu YJ (2018) Liquid crystal enabled dynamic nanodevices. Nanomaterials 8(11). https://doi.org/10.3390/nano8110871
16. Decker M, Kremers C, Minovich A, Staude I, Miroshnichenko AE, Chigrin D, Kivshar YS (2013) Electro-optical switching by liquid-crystal controlled metasurfaces. Opt Express 21(7):8879–8885. https://doi.org/10.1364/OE.21.008879
17. Komar A, Paniagua-Domínguez R, Miroshnichenko A, Yu YF, Kivshar YS, Kuznetsov AI, Neshev D (2018) Dynamic beam switching by liquid crystal tunable dielectric metasurfaces. ACS Photonics 5(5):1742–1748. https://doi.org/10.1021/acsphotonics.7b01343
18. Pryce IM, Aydin K, Kelaita YA, Briggs RM, Atwater HA (2010) Highly strained compliant optical metamaterials with large frequency tunability. Nano Lett 10(10):4222–4227. https://doi.org/10.1021/nl102684x
19. Lapine M, Shadrivov IV, Powell DA, Kivshar YS (2012) Magnetoelastic metamaterials. Nat Mater 11(1):30–33. https://doi.org/10.1038/nmat3168

20. Shih TK, Chen CF, Ho J-R, Chuang FT (2006) Fabrication of PDMS (polydimethylsiloxane) microlens and diffuser using replica molding. Microelectron Eng 83(11):2499–2503. https://doi.org/10.1016/j.mee.2006.05.006
21. Xia Y, McClelland JJ, Gupta R, Qin D, Zhao XM, Sohn LL, Whitesides GM (1997) Replica molding using polymeric materials: a practical step toward nanomanufacturing. Adv Mater 9(2):147–149. https://doi.org/10.1002/adma.19970090211
22. Jahani Y, Arvelo ER, Yesilkoy F, Koshelev K, Cianciaruso C, De Palma M, Altug H (2021) Imaging-based spectrometer-less optofluidic biosensors based on dielectric metasurfaces for detecting extracellular vesicles. Nat Commun 12(1):3246. https://doi.org/10.1038/s41467-021-23257-y
23. Song Q, Zhang W, Wu PC, Zhu W, Shen ZX, Chong PHJ, Liu AQ (2017) Water-resonator-based metasurface: an ultrabroadband and near-unity absorption. Adv Opt Mater 5(8):1601103. https://doi.org/10.1002/adom.201601103
24. Sun S, Yang W, Zhang C, Jing J, Gao Y, Yu X, Xiao S (2018) Real-time tunable colors from microfluidic reconfigurable all-dielectric metasurfaces. ACS Nano 12(3):2151–2159. https://doi.org/10.1021/acsnano.7b07121
25. Guanxing Z, Liu Z, Deng W, Zhu W (2020) Reconfigurable metasurfaces with mechanical actuations: towards flexible and tunable photonic devices. J Opt 23(1):013001. https://doi.org/10.1088/2040-8986/abcc52
26. Zhu WM, Dong B, Song QH, Zhang W, Huang RF, Ting SK, Liu AQ (2014, 26–30 Jan) Tunable meta-fluidic-materials base on multilayered microfluidic system. Paper presented at the 2014 IEEE 27th international conference on micro electro mechanical systems (MEMS)

Chapter 4
Tunable Electromagnetic Resonances with Slab-Split-Ring Meta-molecules

4.1 Introduction

The electromagnetic properties of a material, i.e., permeability and permittivity, are highly dependent on the interaction between the incident electromagnetic waves and the molecules, which typically involve the resonances of electrons in terms of displacements or currents. The metamaterials and metasurfaces respond to the incident electromagnetic waves using the meta-molecules with micro/nano geometric structures where their resonance nature can obtain extraordinary electromagnetic properties. The electromagnetic resonances have long been observed in nature materials at visible frequency regions. For example, intense iridescence can be found when looking at the reflection of a butterfly's wing with multilayered photonic structures [1–5]. Also, the famous Lycurgus Cup shows different colors with different illumination angles due to the metal particles' absorption induced by localized surface plasmon polariton resonances [6–12]. Inspired by those interesting phenomena induced by electromagnetic resonances, many research areas have been established, including micro/nano optical cavities, photonic crystal-based silicon photonics, artificial electromagnetic materials, etc.

Metamaterials and metasurfaces are artificial materials with designable electromagnetic properties due to the resonance natures of their meta-molecules. Despite the flexibility in the design of materials properties, metamaterials and metasurfaces are expected to have electromagnetic responses rare in nature, e.g., negative refractions, near-to-unity absorptions, trivial or super dispersion, etc., which are possible to be realized by controlling the electromagnetic resonances of the meta-molecules. For example, near-to-unity absorption can be realized by confining the electromagnetic fields within the meta-molecules structures with extremely high-quality factors. The dispersion of the meta-molecules can be designed by controlling the multiple electromagnetic resonance modes within their micro/nanostructures. Also, a double-negative refractive index can be achieved by separately tuning magnetic and electric resonances using split-ring and wired structures [13–16]. As a result, vast meta-devices have now been demonstrated using artificial materials with extraordinary

W. Zhu and A.-Q. Liu, *Metasurfaces: Towards Tunable and Reconfigurable Meta-devices*, Microfluidics and Nanophotonics: Science and Engineering 1,
https://doi.org/10.1007/978-981-19-6925-6_4

electromagnetic properties. For example, metasurfaces with optimized permittivity and permeability for optical disguise and electromagnetic cloaking. The subwavelength imaging is achieved by evanescent wave enhancement using metamaterials. Localized electromagnetic field enhancement can also be realized by using metamaterials and metasurfaces. In addition, the breaking of the symmetry of the meta-molecules or their lattice results in the asymmetric resonance modes of the incident electromagnetic waves, which are applied to design artificial materials with significant optical activities. This chapter discusses the electromagnetic resonances within the split-ring meta-molecules as a well-demonstrated example of tunable metasurfaces functionalities.

4.2 Electromagnetic Resonances in Slab-Split-Ring Meta-molecules

The Slab-Split-Ring Meta-molecule

An alternating magnetic flux will induce magnetic resonances when passing through a ring-shaped metal structure. The split ring resonators (SRR) are designed to have strong magnetic resonances at the desired frequency region, typically the same as electrical resonances of the meta-molecules induced by the incident electromagnetic waves. The rod structures have long been used as antennas with electrical dipole resonances. The metamaterial with the meta-molecules composed of a rod and an SRR was firstly proposed by Sir Pendry and then demonstrated by Professor Smith to realize the negative-refractive index material working at the radiofrequency region [17–19]. After that, metasurfaces are proposed with SRR meta-molecules, which show strong electrical and magnetic resonances at the same frequency with proper design.

Here we use a meta-molecule with a slab and an SRR structure to show the structural reconfigurable metasurfaces realized by tuning the magnetic and electrical resonances induced by the incident electromagnetic waves. The SRR is composed of two semi-ring-shaped metal structures, which are noted as "[" and "]". Two metal slabs are placed adjacent to the SRR, as shown in Fig. 4.1. Therefore, the meta-molecule is composed of two metal slabs and an SRR, which is named the slab-split-ring (SSR) meta-molecule. Similar to the flexible substrates discussed in Chap. 2, the metal slabs are patterned on movable silicon frames fabricated using bulk micromachining processes. On the other hand, the SRR metal structures are anchored on the fixed silicon substrate. As a result, the metal slabs can be actuated along the gap direction (x-direction), as shown in Fig. 4.1a. The initial and final states of the meta-molecules are shown in Fig. 4.1b, c, respectively. The electromagnetic properties of the SSR metasurface can be tuned by changing the electrical and magnetic resonances of the meta-molecules. The electromagnetic waves are obliquely incident on the metasurface with altitude angle θ and azimuth angle ϕ. Here, the transverse electric (TE) and transverse magnetic (TM) polarization states are defined by the directions of the electrical and magnetic fields, as shown in Fig. 4.1d, e, respectively.

The electrical field of the TE-polarized incidence is along the x-direction, which is parallel to the metasurface. For TM polarized incidence, the magnetic field is along the x-direction. The metasurface lattice constant along x- and y-directions are both 60 μm, which are denoted as *a*x and *a*y, respectively. The widths of all metal strips are 3 μm, including the metal slabs. The length of the metal slabs and the sidewall of SRR are both 50 μm, while the gaps of the SRR are 20 μm. The distance between the metal slabs and the SRR sidewall is 2 μm.

Electromagnetic Resonances in SRRs

Here, the electromagnetic resonances with a split ring resonator are discussed to illustrate the working principle of the SSR metasurfaces. The numerical simulation results are obtained by using the finite-difference time-domain (FDTD) method discussed in Chap. 1. The homogeneous metasurface with SRR meta-molecules is studied with oblique incidence with both TE and TM polarized light, as shown in Fig. 4.1. Both magnetic and electrical resonances can be analyzed by using the metasurface's reflection spectra and surface current. The incident angles of the electromagnetic waves are defined by the altitude angle θ and azimuth angle ϕ, as illustrated in Fig. 4.1a. Figure 4.2 shows the reflection spectra of the SRR metasurface and the surface currents distribution at the resonance frequencies. The metasurface works in the THz region, while the calculated spectra range from 0.5 to 1.2 THz. The azimuth angle ϕ is fixed at 0° for both TE and TM polarized light so that the incident plane is perpendicular to the x-direction. The SRR structure is composed of aluminum with a conductivity of 3.72×10^7 S/m. The reflection spectra are simulated under different altitude angles θ ranging from 0° to 45°.

The periodic envelopes of the reflection spectra with a free spectra range (FSR) of 0.38 THz are observed due to the Fabry-Pérot (FP) effect between the metal layer and the backside of the silicon substrate. The FSR of the silicon substrate can be calculated by using the equation FSR = c/2n*d*, where c = 3 × 108 m/s is the free-space velocity of the electromagnetic waves, n = 3.45 is the refractive index of silicon at the THz region, and d = 100 μm is the thickness of the silicon substrate. It should be pointed out that the calculated FSR, which is 0.43 THz in this case, may not be the same as observed in the reflection spectra, which is due to the phase variation induced by the metal structures of the meta-molecules. A more accurate method of calculating the FSR should consider the phase of reflected electromagnetic waves from the meta-molecules.

Figure 4.2 shows a resonance dip at 0.88 THz for TE polarized incidence under different altitude angles θ, which can be analyzed by the surface current distributions, as shown in the inserts. The electrical and magnetic fields are parallel to the metasurface at normal incidence when $\theta = 0°$. There is no magnetic flux perpendicular to the SRR surface to induce magnetic resonances. The projection of the magnetic field along the z-direction increases for TE polarized incidence increases when the altitude angle θ is increasing. On the other hand, the electrical field of the incidence remains the same when the altitude angle θ changes. It can be observed from Fig. 4.2a

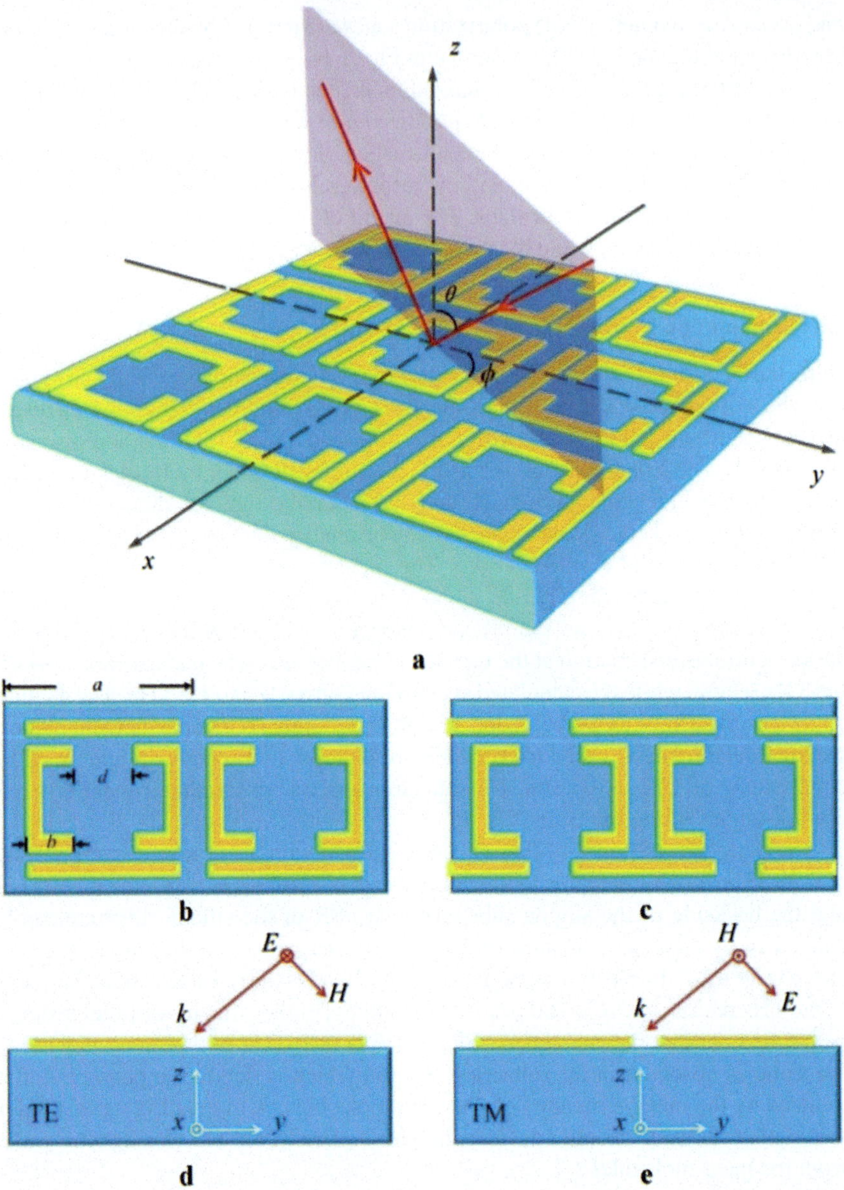

Fig. 4.1 Schematics of the reconfigurable metasurface with slab-split-ring meta-molecules **a** perspective review of the metasurfaces and incident conditions **b** and **c** the initial and final states of the slab-split-ring meta-molecules, respectively **d** and **e** the definitions of the transverse electric (TE) and transverse magnetic (TM) polarization states, respectively

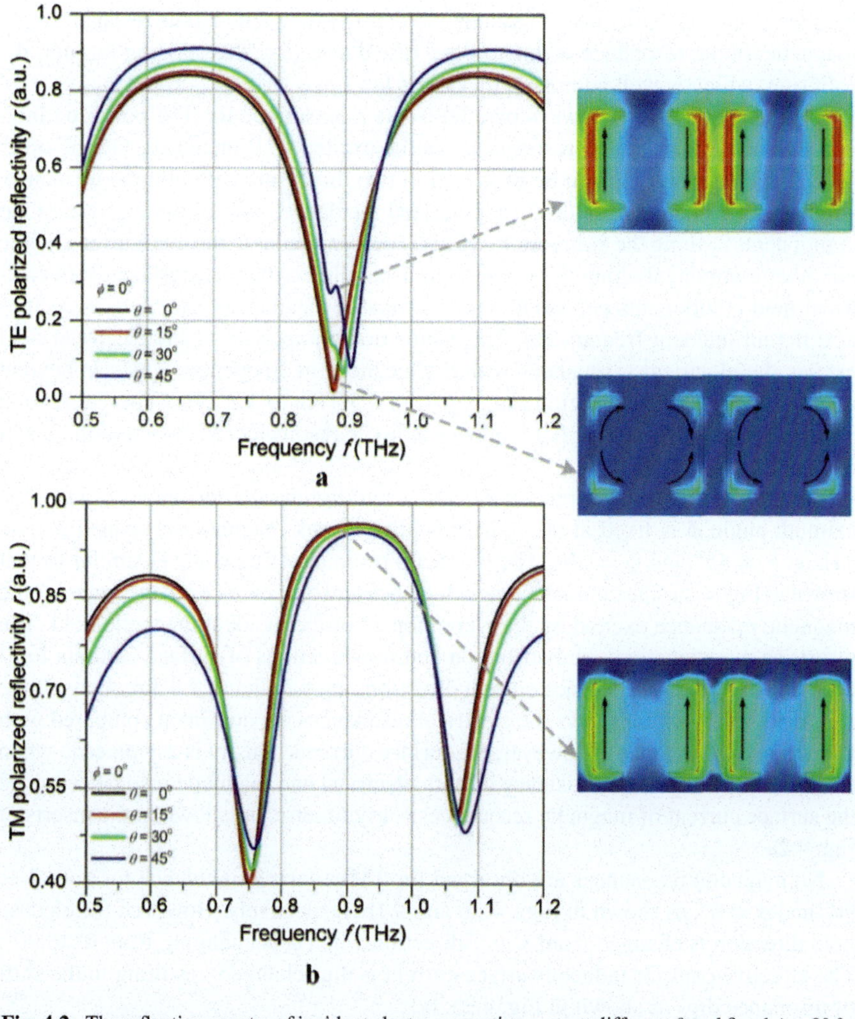

Fig. 4.2 The reflection spectra of incident electromagnetic waves at different θ and fixed $\phi = 0°$ for **a** TE and **b** TM polarization states: The inserts show surface current at $\theta = 0°$ and $\theta = 45°$ at 0.88 THz for TE polarized incidence and $\theta = 0°$ at 0.88 THz for TM polarized incidence, respectively

that a small resonance peak at 0.88 THz is induced by the variation of the incident magnetic field.

The surface current distributions at $\theta = 0°$ and $\theta = 45°$ are analyzed to further investigate the induced magnetic resonance at 0.88 THz incident frequency. The inserts of Fig. 4.2 show the surface currents distribution of two adjacent SRR meta-molecules. The arrows represent the instantaneous direction of the surface current, and the color indicates the amplitude of the magnetic fields. The surface current on both "[" and "]" flow in the x-direction, indicating an electrical dipole resonance when $\theta = 0°$. The magnetic field amplitude is much smaller than that of $\theta = 45°$ when the surface currents direction is circular, indicating a magnetic dipole resonance. The magnetic dipole resonance is due to time-varied magnetic flux passing through the split ring, which is well addressed by Lenz's law.

The magnetic field is always parallel to the metasurface for TM-polarized incidence, where no magnetic resonances can be excited as θ increases. On the other hand, the electrical field can be projected in two directions with oblique incidence, similar to the magnetic field of TE-polarized incidence. One of the electrical field components is along the z-direction, which is perpendicular to the metasurface, while the other component along the y-direction is parallel to the metasurface. Both electrical field components can excite the electrical resonances of the meta-molecules at different incident frequencies. The z-direction component of the electrical field excites the electrical resonances with higher incident frequency, which is beyond the discussed frequency region ranging from 0.5 THz to 1.2 THz. The y-direction component of the electrical field induces a dipole resonance, as shown in the insert of Fig. 4.2b.

Figure 4.3a, b show the reflection spectra at different altitude angles θ when the azimuth angle ϕ is fixed at 90°. There is also a weak magnetic resonance at 0.89 THz at $\theta = 45°$ and $\phi = 90°$. The magnetic resonance frequency is similar to that shown in Fig. 4.2a since the effective inductance and capacitance are the same for the magnetic resonance excited by the z-component of the incident magnetic field. The inserts show magnetic field distribution and the directions of surface currents for $\theta = 0°$ and $\theta = 45°$. The surface current directions are parallel to the directions of the electrical fields, which shows the electrical resonances are dominant compared with the magnetic resonances. However, the surface currents density is asymmetric when $\theta = 45°$ indicating the coexistence of both electrical and magnetic resonances since the surface current of magnetic resonances is asymmetric, as shown in the inserts of Fig. 4.2a.

No magnetic resonances are observed for TM-polarized incidence for both $\phi = 90°$ and $\phi = 0°$, as shown in Figs. 4.2b and 4.1b, respectively. However, the electric field direction is changed from x to y-direction when ϕ is changing from 0° to 90°. The effective path for induced surface current is thus changed, resulting in the shift of resonance dips as shown in Fig. 4.2a, b.

The reflection spectra with different SRR gaps are analyzed to show the tuning of the magnetic resonances with the meta-molecule reconfiguration. Here, only TE-polarized incidence is discussed since no magnetic resonances are observed for TM-polarized incidence. Figure 4.3d shows the reflection spectra with TE-polarized incidence for different SRR gaps. The altitude angle θ is fixed at 45° when the gaps increase from 5 to 20 μm. The top and bottom graphs show the reflection spectra when $\phi = 90°$ and $\phi = 0°$, respectively. Both graphs show that the magnetic resonance frequencies vary from 0.82 to 0.89 THz when the split gap increases.

It can be concluded that both TE and TM-polarized incidence can induce electrical resonances. In contrast, magnetic resonances can only be induced by TE-polarized incidence with a magnetic field perpendicular to the SRR structure at oblique incidence. The magnetic resonances are much weaker than the electrical resonances for the SRR metasurface, mainly due to the single-layered structure. The electrical

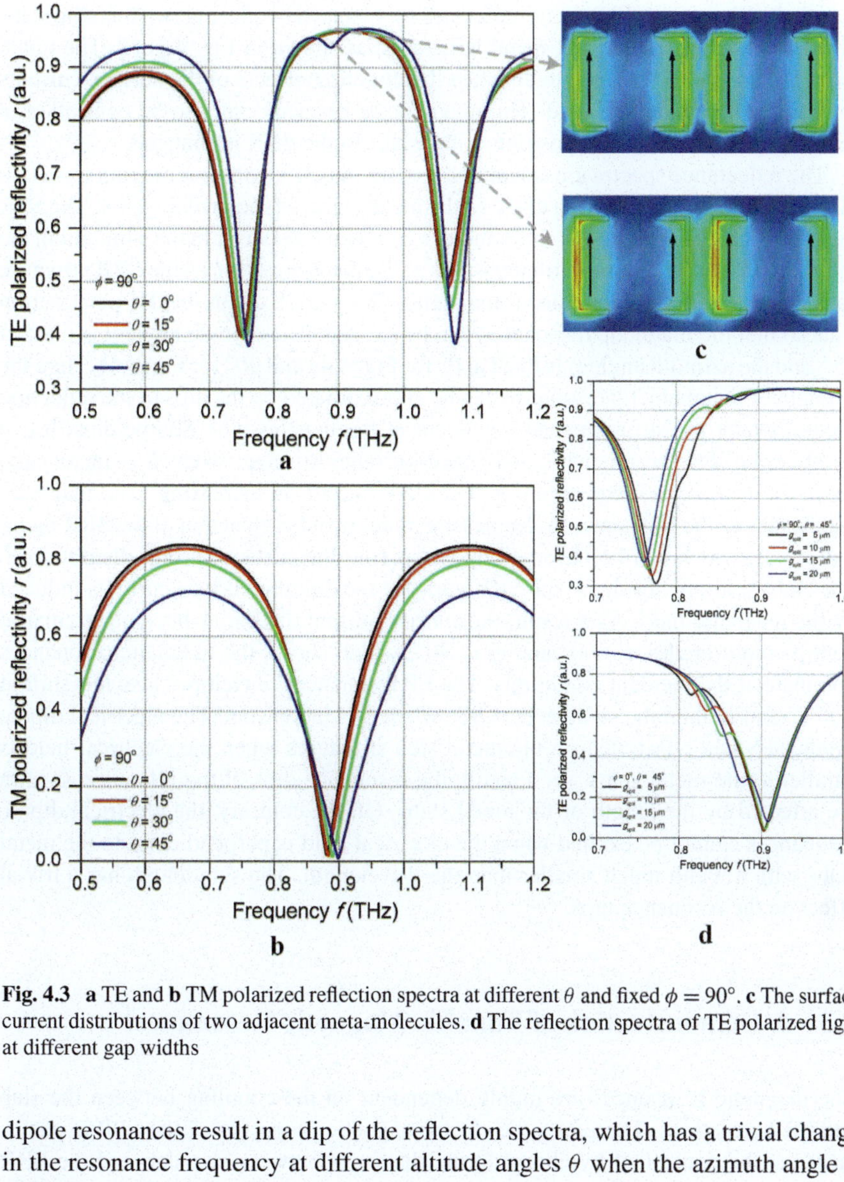

Fig. 4.3 **a** TE and **b** TM polarized reflection spectra at different θ and fixed $\phi = 90°$. **c** The surface current distributions of two adjacent meta-molecules. **d** The reflection spectra of TE polarized light at different gap widths

dipole resonances result in a dip of the reflection spectra, which has a trivial change in the resonance frequency at different altitude angles θ when the azimuth angle ϕ is fixed.

The Magnetic Resonance Enhancement Due to the Slabs

The SRR meta-molecules mentioned above have magnetic resonances at the THz range induced by TE-polarized electromagnetic waves with oblique incident angles. However, the magnetic resonances of the SRR are very weak compared with the electrical resonances due to the near-field coupling between the structure "[" and structure "]". The magnetic field is not well confined in the meta-molecules. The coupling effect of the SRR meta-molecules is enhanced by placing two metallic slabs beside both sides of the SRR metal structures, as shown in Fig. 4.1b, c. The metal slabs are used as a "bridge" to enhance the coupling effects of the surface currents induced by the magnetic field. The magnetic resonances can also be controlled in real-time when the relative position of the slabs to the SRR is changed.

The reflectance spectra are calculated at different slab lengths L ranging from 0 to 50 μm to analyze the coupling effect of the metal slabs, as shown in Fig. 4.4. The SSR meta-molecules become SRR meta-molecules when $L = 0$ μm. In the simulation, the metal slabs share the same symmetry axis along the y-direction as the SRR structure as L changes. The incident electromagnetic waves are fixed at the TE-polarization state so that the magnetic resonances can be excited. The altitude angle θ is fixed at 45°, and the azimuth angle ϕ is fixed at 0° for Fig. 4.4a and 90° for Fig. 4.4b. Here the strength of the magnetic resonance can be approximated by the dips of the reflection spectra where the strong resonances occur at frequencies with Sharpe dips. It can be observed that the magnetic resonance becomes stronger when L is increasing. Also, the resonance frequency is red-shifted when L is increasing. The magnetic resonances of the SSR meta-molecules are much stronger than that of the SRR meta-molecules, which can be observed by comparing the surface current distributions. The slab structure "bridges" the SRR gaps for the circular surface currents induced by the magnetic field. As a result, the surface current directions become circular at both azimuth angles $\phi = 0°$ and $\phi = 90°$, which shows the magnetic resonances dominate at the incident conditions. The FP resonances' envelopes also red-shift at $\phi = 0°$ and remain the same at $\phi = 90°$ when L is increasing. The electrical dipole resonances can be excited on the metal slab structures when the electrical field is parallel to the metal slabs. As a result, the resonance dips of the reflection spectra are affected by the length of the metal slabs. On the contrary, the electrical dipole resonances cannot be excited when the electrical field is perpendicular to the metal slabs with a width much smaller than the wavelength. The slab length has a trivial effect on the resonance dips.

4.3 A Demonstration of Tunable Magnetic Resonances

The magnetic resonances are highly dependent on the coupling between the slab and SRR structures. Therefore, the electromagnetic properties of the SSR metasurfaces can be controlled by changing the distances between the slab and the SRR structures. A straightforward method is to tune the coupling strength by adjusting the gap width between the SRRs and the slabs, significantly changing the magnetic resonance frequency and strength. However, this method requires a mechanical actuation along the y-direction, as shown in Fig. 4.1, which is quite limited by the narrow spacing between adjacent SSR meta-molecules. Additionally, the actuation along the y-direction cannot show a limited effect for the electrical resonances when the incident electric field is along the x-direction. Therefore, the actuation of the slab is chosen to be along the x-direction.

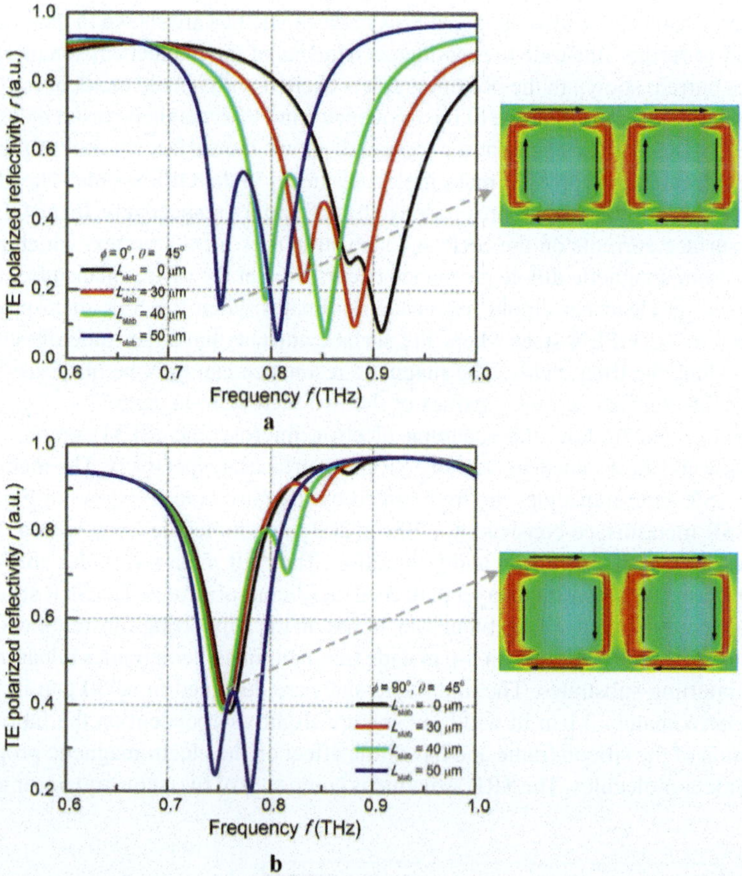

Fig. 4.4 The refection spectra of TE-polarized incidence at $\theta = 45°$and **a** $\phi = 0°$ and **b** $\phi = 90°$. The inserts show surface current on the SSR structure at 0.75 THz

Here, the lattice constant of the SSR meta-molecules is 60 μm along the x-direction. The actuation distances for the slabs are capped at 30 μm due to the translation symmetry of the SSR meta-molecules along the x-direction. Figure 4.5 shows the reflection spectra of the SSR metasurface with different slab displacement Δs. The CLOSE-state of SSR meta-molecules is defined as $\Delta s = 0$ μm, while the OPEN-state is defined as $\Delta s = 30$ μm, as shown in Fig. 4.1b, c, respectively. Here the CLOSE and OPEN are used to describe the geometry of the SSR meta-molecules when the SRR gaps are completely closed or opened by the slab structures. At the OPEN-state, the magnetic resonance frequency is 0.77 THz, as shown in Fig. 4.5. The actuation of the slab structure along the x-direction dramatically decreases the coupling between the two semi-squares of the SRR structures, which also results in a blue shift of the magnetic resonance frequency. The magnetic resonance peaks vanish when the displacement of the slab Δs is larger than 20 μm. The surface current distributions of two adjacent SSR meta-molecules are shown in the inserts of Fig. 4.5 to further illustrate the tuning mechanisms of the magnetic resonances. The surface current shown in the bottom graph is excited with a fixed incident frequency of 0.74 THz, which is the edge of the frequency band with magnetic resonances at the CLOSE-state. For comparison, the top graph shows the surface current of the same incident frequency with SSR meta-molecules tuned to the OPEN-state. The altitude angle θ and azimuth angle ϕ are chosen to be 45° and 0°, respectively. In the CLOSE-state, surface currents on the SRR and slab structures form a vortex, which proves that a strong magnetic dipole resonance occurs even at the off-resonance frequency. However, an electrical dipole resonance dominates when the SSR meta-molecule is tuned to the OPEN-state where the surface currents have the same direction as the incident electrical fields. The magnetic resonance can only be observed by the asymmetric surface current densities of the two SRR semi-squares.

Figure 4.6a, b show the scanning electron microscopic (SEM) graphs of the overview and the zoomed-in view of SSR metasurfaces, respectively. The metal parts of the SSR meta-molecules are highlighted by the false color (yellow) in Fig. 4.6b. The SSR metasurface consists of a 200 × 200 meta-molecule array with a 60-μm period along both x-direction and y-direction. The SSR meta-molecules are formed by a 0.5-μm thick aluminum layer patterned on silicon substrates. The SRR structures are patterned on fixed silicon blocks anchored on the supporting substrate, while the slab structures are patterned on 3-μm wide free-hanging silicon frames released from the supporting substrates. The aluminum slabs are designed to be 50 μm in length and approximately 3 μm in width, which are slightly dependent on the fabrication condition of the silicon frame and have little effect on the electromagnetic properties of the meta-molecules. The SRR structure is composed of two semi-square rings with

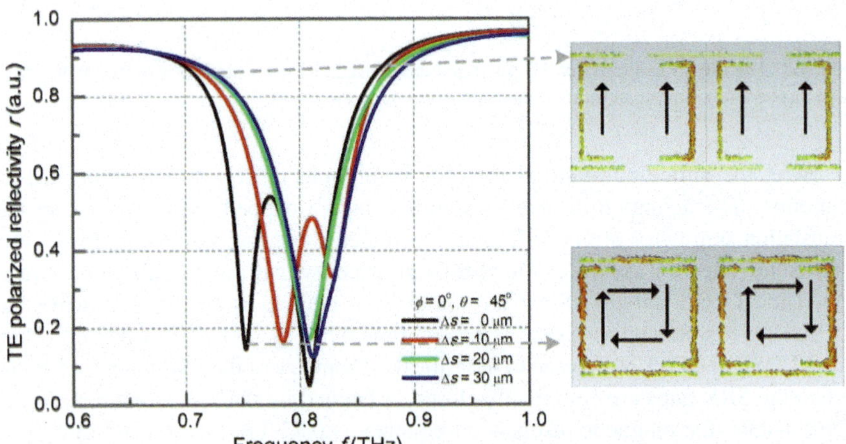

Fig. 4.5 The refection spectra of SSR metasurface for different slab displacements with TE-polarized incidence at $\theta = 45°$and $\phi = 0°$. The inserts show surface current on the SSR structure at 0.75 THz

a 50-μm sidewall along the x-direction and a 20-μm sidewall along the y-direction, which are patterned on silicon blocks much larger than the silicon frames of the slab structures. As a result, the silicon frames are released from the supporting substrate, while the silicon blocks are not after the same releasing process. Two micromachined comb-drive actuators actuate the slab structures with bidirectional in-plane actuation along the y-direction.

Figure 4.6c shows the photograph and schematic of the experimental setup within commercial FTIR equipment (Bruker Vertex 80v). A blackbody source is used for the THz spectra measurement, while a mercury cadmium telluride (HgCdTe) detector cooled with liquid nitrogen is used for signal detection. The reflection spectra measurement setup consists of four mirrors, denoted as M1 to M4, as shown in Fig. 4.6c. The incident light is collected by M1 and collimated using a curved mirror M2. Then the collimated THz waves are incident on the SSR metasurface while M3 and M4 collect the reflected light. Here, the incident angle can be tuned by rotating M1 and M2, where M3 and M4 are rotated accordingly to collect the reflected THz

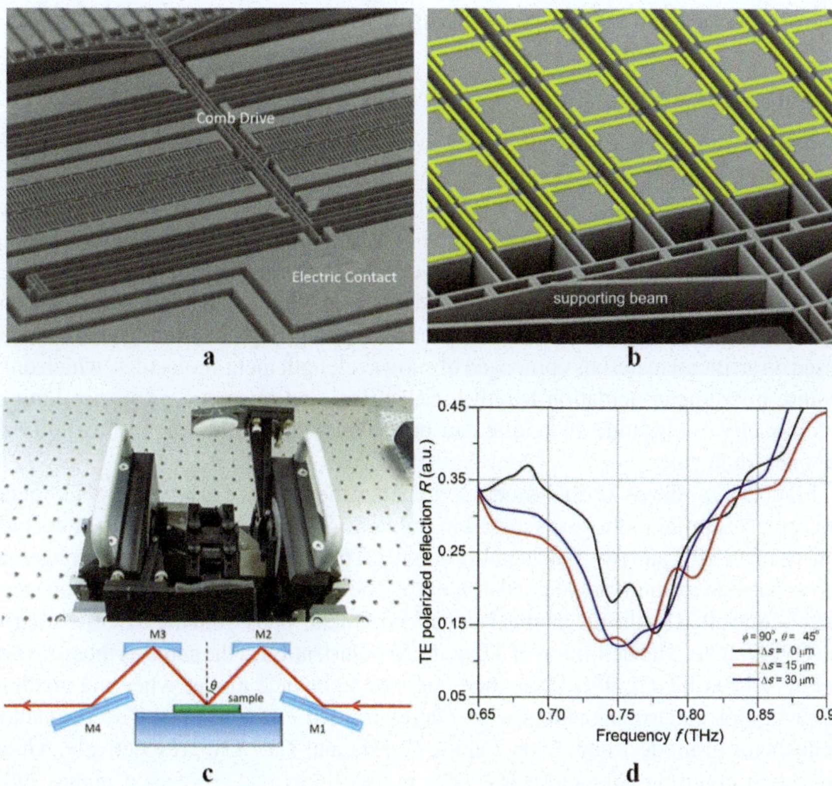

Fig. 4.6 **a** Overview and **b** zoomed-in views of SEM graphs of the fabricated SSR metasurfaces. **c** The experimental setup and **d** the measured reflection spectra of TE-polarized incidence at $\theta = 45°$ and $\phi = 90°$

waves from the metasurface. In the experiment, the incident angle can be tuned from 15° to 75°.

Figure 4.6d shows the reflection spectra of the SSR metasurface measured from 0.65 THz to 0.90 THz, where the magnetic resonances are dominating. Here the altitude angle θ is fixed at 45°, and the azimuth angle ϕ is fixed at 90°. The wide dips of the reflection spectra are due to the FP resonances, while the narrow resonance peaks are due to the magnetic resonances. The displacement of the slab structures tunes the resonance peaks. In Fig. 4.6d, the reflection spectra are measured when $\Delta s = 0$ μm (CLOSE-state), 15 μm, and 30 μm (OPEN-state). The displacement is realized by applying electric voltages on the electrostatic comb-drive actuators, which are 0 V, 20 V, and 27 V, respectively. The displacement and actuation voltage are calibrated before the experiment so that Δs is obtained by the voltage applied to the comb drive actuator. It can be observed that the magnetic resonances are significantly suppressed when the SSR is turned from CLOSE-state to OPEN-state.

4.4 Resonance Modes Switching and Tuning

Dual Mode Switching

Switchable electromagnetic devices have been widely demonstrated based on mechanical actuation or nonlinear properties of the materials, e.g., the thermo-optical effect, free carrier effect, photon lumination effect, etc. Most fast switching electromagnetic devices are based on tuning the electrical and magnetic resonances using nonlinear materials, the switching speed of which can go up to THz. On the other hand, the mechanical actuation is slow and unstable with movable parts. Now reconfigurable metasurfaces offer a new approach to designing electromagnetic devices based on artificial materials composed of subwavelength meta-molecules, which only require micrometer actuation length for electrical and magnetic resonance tuning. Additionally, multimode switching can be realized due to the design flexibility of the meta-molecules.

Figure 4.7a shows a dual-mode switchable metasurface using the SSR meta-molecules. Similar to those discussed in previous sections, the SSR meta-molecules consist of an SRR and two slab metal structures. The oblique incident electromagnetic waves have an azimuth angle fixed at $\phi = 0°$. The electrical and magnetic resonances can be controlled by the mechanical actuation of metal slabs from the CLOSE-state to the OPEN-state. The definition of TE and TM polarization is the same as those shown in Fig. 4.1. At the CLOSE-state, there are two resonance modes when the incident frequency of electromagnetic waves ranges from 0.7 to 1.2 THz. The resonance frequencies of mode 1 and mode 2 are 0.75 THz and 1.15 THz, respectively, which share two common characteristics. One is that these two resonance modes only appear with oblique TE-polarized incidence where $\theta > 0$. The other is that these two resonance modes disappear when the meta-molecules are tuned from the CLOSE-state to the OPEN-state. Therefore, both mode 1 and mode 2 are dominated by magnetic resonances.

The surface current distributions of the two magnetic resonance modes are shown in Fig. 4.7b, c for CLOSE-state and OPEN-state, respectively. The red arrows show the instantaneous directions of the surface currents, while the density of the arrows on the metal parts represents the intensity. At the CLOSE-state, the surface current forms a circular loop indicating a strong magnetic dipole resonance of mode 1, discussed in previous sections. However, mode 2 has unbalanced surface currents at the two

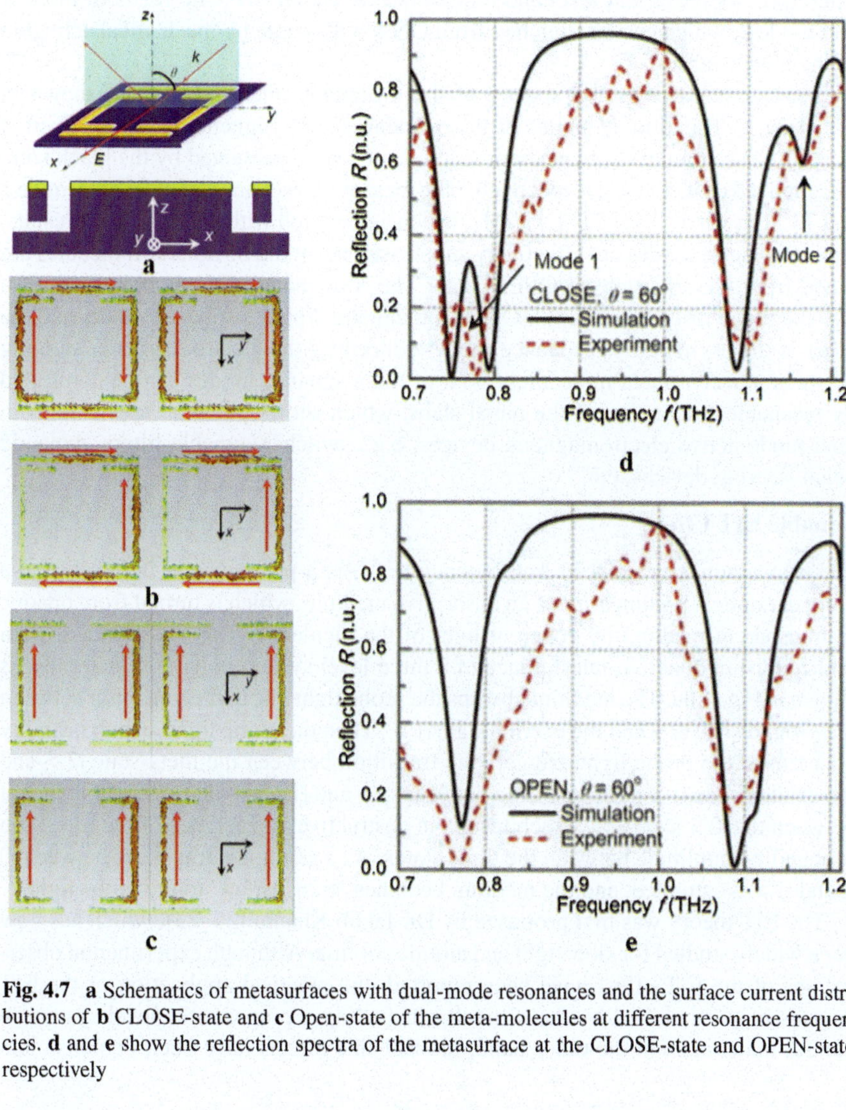

Fig. 4.7 **a** Schematics of metasurfaces with dual-mode resonances and the surface current distributions of **b** CLOSE-state and **c** Open-state of the meta-molecules at different resonance frequencies. **d** and **e** show the reflection spectra of the metasurface at the CLOSE-state and OPEN-state, respectively

semi-square structures, as shown in the bottom graph of Fig. 4.7b. More interestingly, the direction of the surface current on the SRR structure is reversed compared with mode 1, which shows that the electrical resonance dominates in mode 2. Dipole resonances can be excited by the electrical fields on the SRR structures along the x-direction. However, the electrical field incident phases on the two semi-squares are different due to the oblique incident. The surface current on the left semi-square structure is suppressed by the destructive interference of the meta-molecules, which shows a dark resonance mode. At the OPEN-state, no circular currents are observed for both mode 1 and mode 2. The coupling between the SRR and slab structures still exists, which is much weaker than that of the meta-molecule at the CLOSE-state. Therefore, the electrical resonances dominate at the OPEN-state for both mode 1 and mode 2, which can be switched from ON to OFF-state by mechanical actuation of the metal slabs.

The measured reflection spectra of the switchable metasurfaces are shown in Fig. 4.7d, e. The altitude angles of the incident electromagnetic waves are fixed at $\theta = 60°$. The simulation and experimental results are represented by the black solid lines and red dashed lines, respectively. The measured resonance frequency of mode 1 is 0.73 THz, while mode 2 is 1.17 THz, which agrees well with the simulation results. Both resonance modes are effectively suppressed when the SSR meta-molecules are tuned from CLOSE-state to OPEN-state. The SSR metasurface is proven to have two magnetic resonance modes. One is a magnetic dipole resonance mode, and the other is the asymmetric resonance mode induced by the electromagnetic coupling between adjacent meta-molecules. Both of the resonance modes can be controlled by mechanical actuation of the metal slabs, which shows potential applications in dual-mode active electromagnetic devices, e.g., switches, tunable filters, dynamic beam steering devices, etc.

Tunable EIT Effect

Electromagnetically induced transparency (EIT) is a phenomenon induced by the coherent optical nonlinearity of an absorptive medium, which is turned from opaque to transparent medium by coherent light of the incidence. The EIT effect is often realized using quantum interference of a three-level system in natural materials, as shown in Fig. 4.8a. The medium absorbs the probe light due to the transition between the ground state $|0>$ and the excited state $|1>$. Then the pump light, which is highly coherent to the probe light, triggers the transition between the meta-state $|2>$ and the excited state $|1>$. The coupling between $|2>$ and $|1>$ is much stronger than that between the $|0>$ and $|1>$, which results in destructive interference of the transition probability amplitude between the three states. As a result, the transition between $|0>$ and $|1>$ is saturated, and the medium becomes "transparent" to the probe light.

The EIT theory was first proposed by Dr. Jakob Khanin [20–23], which has now been widely studied for slow light and atomic cooling. Although experimental observations of the EIT effect have been reported, the practical application of the EIT devices is quite limited by the difficulty of finding a proper three-level system in natural materials. On the other hand, the EIT-like effects have been demonstrated

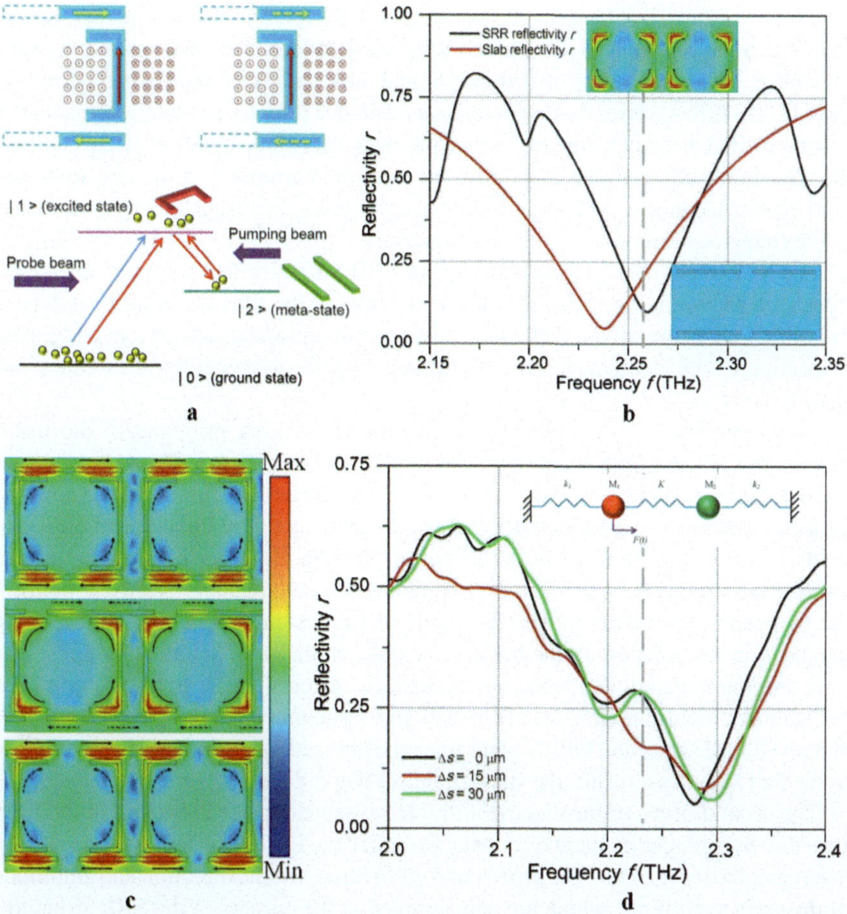

Fig. 4.8 **a** Schematics of the EIT-like effect induced by the SSR metasurface, **b** the reflectivity spectra of SRR structures and metal slabs, **c** and **d** show the measured transmissivity and reflectivity spectra of SSR metasurfaces, respectively

using micro-cavities with strong near-field couplings. Later, the EIT effect was introduced to design meta-molecules for metasurface devices based on extraordinary electromagnetic properties.

Generally speaking, only electrical resonances can be excited within the metasurfaces by normal incidence due to the 2D structures of the metasurfaces. However, the coupling between the electrical and magnetic resonances is non-essential for realizing the EIT-like metasurfaces. The three-level system can also be realized by the coupling between multiple electrical resonance modes. On the other hand, the coupling between electrical and magnetic resonances is challenging to be realized in a 2D metasurface. Therefore, the normal incidence is chosen to demonstrate EIT-like metasurfaces based on the coupling between multiple electrical resonance modes, which can be tuned by changing the geometries of the meta-molecules. In this section, only electrical resonances are excited with the EIT-like metasurfaces with normal incidence. The coexistence of both bright and dark resonance modes is required for the EIT-like effect. Here, the bright resonance modes refer to the electrical resonance modes excited directly by the free-space incident electromagnetic waves. In contrast, the dark mode can only be excited by the near-field coupling of the bright modes. Dark resonance modes typically result in a high-quality factor (Q factor), asymmetric in the frequency domain, e.g., Fano resonances.

The SSR meta-molecules consist of an SRR structure and two movable slabs, which are shown in Fig. 4.8a. The lattice constant of the SSR metasurface is 60 μm in both x and y-directions, while the slabs' length and width are 50 μm and 2 μm, respectively. The incident electrical fields are along the longer sidewalls of the semi-square structures.

The graph on the top of Fig. 4.8a shows the working principle of the meta-molecules at the CLOSE-state (left) and the OPEN-state (right). The yellow arrows indicate the instantaneous directions of the surface currents on the slab structures. The surface currents on the SRR structures are driven directly by the incident electrical fields, which is illustrated by the red arrows in Fig. 4.8a. However, the surface currents on the slab structures cannot be directly induced by the incident electrical fields in the perpendicular direction. The dot-circle and cross-circle symbols represent the magnetic fields induced by the surface currents on the semi-square structures. The time-varying magnetic fields induce the surface currents on the slab structures. The net magnetic fields are non-zero for a half slab structure when the meta-molecule is at both CLOSE-state and OPEN-state since the induced magnetic field is asymmetric to the half slab. As a result, the slab structures can induce magnetic resonances.

The second order of the electrical dipole resonance mode is excited in the SRR structure by a proper incident frequency to realize the EIT-like effect, as shown in the insert graphs of Fig. 4.8b. The arrows and color represent the direction and amplitude of the surface currents, which are concentrated at the corners of the SRR structures for strong coupling between the SRR and slab structures. Figure 4.8b shows the reflection spectra from 2.15 to 2.35 THz for both SRR and slab structures, which are represented by the black and red lines, respectively. The reflection of the SRR structures is much higher than that of the slab structures at 2.21 THz, which shows that the electrical resonance mode of SRR structures has stronger coupling to the reflection mode for the free-space propagation.

Here, the EIT-like resonance mode interference within the SSR meta-molecule is designed by coupling the electrical resonances between the SRR and slab structures. The ground state $|0>$, the excited state $|1>$ and the meta state $|2>$ are represented by the SRR resonance mode, free-space reflection mode, and slab resonance mode. The coupling between the SRR resonance mode and the free-space reflection mode results in the reflection dip at 2.2 THz of the SRR refection spectrum, which is an analogy to the transition from quantum states $|0>$ to $|1>$. The metasurface shows a strong absorption of the reflection. However, the slab resonance suppresses the coupling between the SRR resonance and the free-space reflection, which results in a higher reflection.

The reflection spectra of metasurfaces with meta-molecules composed of only SRR and slab structures are shown in Fig. 4.8b. The refection dip of the SRR metasurface is at 2.26 THz, indicating an absorption band of the metasurface. The surface currents distributions of SSR meta-molecules are shown in Fig. 4.8c. The top, middle, and bottom graphs show the surface currents of two adjacent SSR meta-molecules tuned at the CLOSE-state $\Delta s = 0\ \mu m$, the MIDDLE-state $\Delta s = 15\ \mu m$, and the OPEN-state $\Delta s = 30\ \mu m$, respectively. The amplitudes and directions of the surface currents are represented by the color and the arrows, respectively. The surface currents are concentrated at the slab structures when the meta-molecules are tuned to the CLOSE-state or the OPEN-state. At the MIDDLE-state, the surface currents are mainly concentrated at the SRR structures, similar to the meta-molecules composed of SRR structures only.

Figure 4.8d shows the measured reflection of the EIT-like metasurface when the meta-molecules are tuned to the CLOSE-state $\Delta s = 0\ \mu m$, the MIDDLE-state $\Delta s = 15\ \mu m$, and the OPEN-state $\Delta s = 30\ \mu m$. The experimental setup is similar to that as shown in Fig. 4.6c, which cannot measure the reflection of the normal incidence since the incident and reflected THz waves cannot be separated. Therefore, a small altitude angle with $\theta = 15°$ is used for the reflection spectra measurement. The FP resonance results in a Gauss-like envelope with a period of about 0.4 THz. A reflection peak due to the EIT-like effect is observed at 2.22 THz when the meta-molecules are tuned to the CLOSE-state and the OPEN-state, as shown by the green and black solid lines. The reflection peaks are suppressed when the meta-molecules are tuned to the MIDDLE-state, indicating a high absorption of the SSR metasurfaces. As a result, dynamic control of the EIT-like effect can be realized by the mechanical actuation of the slab structures. The difference between the resonance frequencies of the simulation and experimental results is mainly due to the tilted incidence required by the experimental setup.

Fano Resonance Reconfiguration

Ettore Majorana first observed the asymmetric line shape of inelastic scatting of the electrons from helium. In 1967 Italian-American physicist Ugo Fano gave the theoretical explanation of this wave phenomenon, which is named Fano resonance. The asymmetric line shape of the spectrum results from the interference between a background and a resonant scattering process, which was later demonstrated by the coupling between a high-Q and a low Q resonance. In this section, SSR meta-molecules are designed to realize the dynamic control of the electrical resonance modes, which are switched between dipole modes to quadrupole modes. The polarization states of the normal-incident THz waves are defined by the orientation directions of the electrical fields, which are x- and y-polarized incidence when the incident electrical fields are along x- and y-directions, respectively. The dynamic reconfiguration of Fano resonances has promising applications on fast mode-switching meta-devices working in the THz region.

The SSR meta-molecules consist of an SRR structure and two slabs, which are the same as shown in Fig. 4.7a. The electrical dipole resonance modes can be excited by the normal incidence on both SRR and slab structures. The x-polarized incidence induces the dipole resonance in the SRR structure, while the y-polarized incidence excites the dipole resonance in the slab structures. As discussed in previous sections, the electrical dipole resonances couple to the reflective modes of the free-space propagating light, which results in a high-Q resonance dip in the transmission spectra. On the other hand, the Q factor of the FP resonance induced by the substrate is lower than those of the electrical dipole resonances. As a result, Fano resonances are induced by the interference between the high-Q electrical dipole resonances and low-Q FP resonances.

Figure 4.9a, b show the transmission spectra of SRR structures and slab structures, respectively. Both of them are anchored on a silicon substrate of 100 μm in thickness. The red-dashed and black-solid lines represent the transmission spectra under y- and x-polarized incidence, respectively. The inserts show the schematics of the SRR and slab structures. The blue-solid line shows the transmission spectrum of a 100-μm silicon substrate. The SRR structures show a strong electrical dipole resonance under x-polarized incidence, which results in a transmission dip at 2.6 THz. However, the electrical dipole resonances of the SRR structures become weak due to the existence of gaps between the semi-square structures. As a result, the Fano resonances of SRR structures induced by y-polarized incidence have a transmission dip at 2.55 THz, which is much weaker than those induced by x-polarized incidence. No Fano resonances can be observed within the slab structures under x-polarized incidence. The surface currents distributions on both SRR and slab structures are shown in Fig. 4.9c, d under x- and y-polarized incidence. The color and arrows, respectively, represent the amplitudes and directions of the surface currents. Strong dipole resonances are observed within the SRR structure, while no dipole resonances are found in the slab. The SRR and slab structures show the dipole resonances under y-polarized incidence.

The Fano resonances can be tuned by the mechanical actuation of the slab structures. Figure 4.9e, f show the transmission spectra of the SSR metasurfaces with different slab displacement Δs under x- and y-polarized incidence, respectively. The Low-Q FP resonances induced by the silicon substrate result in the Gaussian-shaped envelopes of the transmission spectra. The interferences between electrical dipole resonances and the FP resonances result in the resonance dips of the transmission spectra near 2.6 THz. The depth of the resonance dips can estimate the Fano resonance strength. The Fano resonance strength induced by the x-polarized incidence increases and then decreases when the meta-molecules are tuned from the CLOSED-state to OPEN-state. On the other hand, the Fano resonances are suppressed at the MIDDLE-state $\Delta s = 15$ μm when the incident THz waves are y-polarized.

The tuning of the Fano resonances can be explained by the mode coupling effects under different polarization states. The x-polarized THz waves only induce electrical dipole resonances on the sidewalls of the SRR structures, which couple to the slab structures and form a quadrupole resonance mode. The coupling strength between dipole and quadrupole modes depends on the symmetry of the SSR meta-molecules. The incident x-polarized THz waves couple to the free-space transmission and reflection modes via quadrupole and dipole resonances. As a result, the Fano resonance

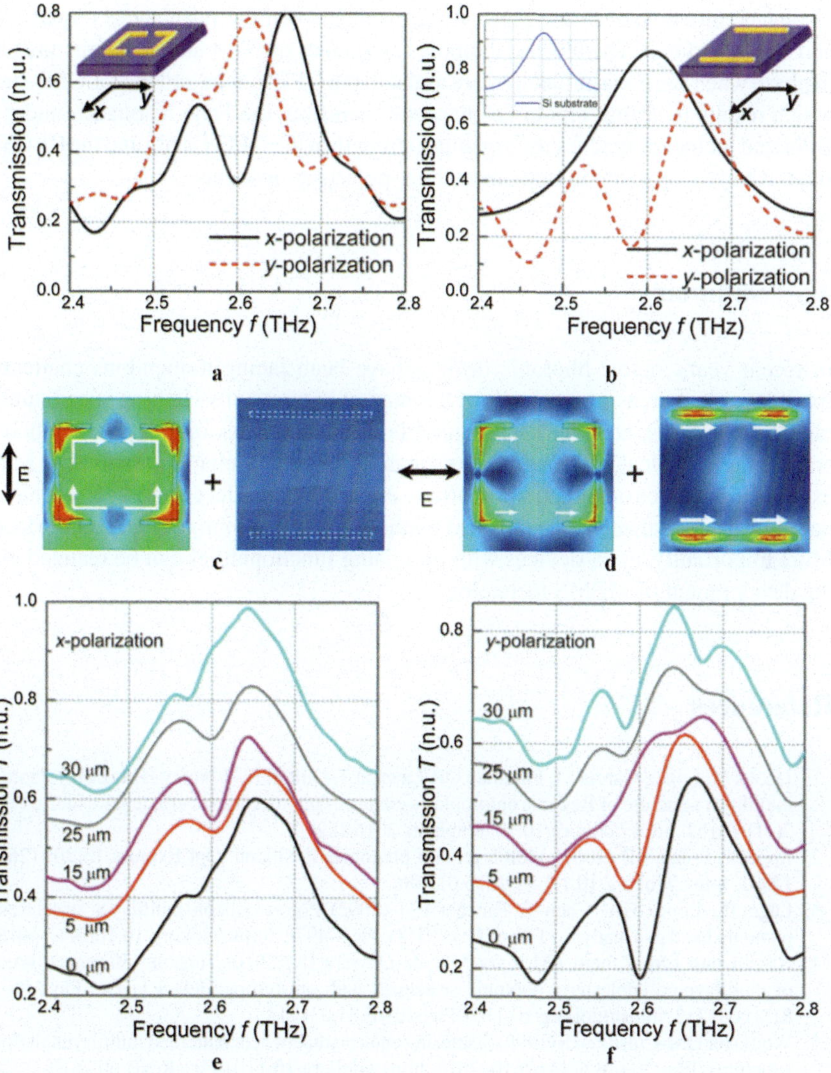

Fig. 4.9 Transmission spectra of **a** SRR structure and **b** metal slabs with different incident polarization states and the surface current distributions of SRR and slab structures with **c** x-polarized and **d** y-polarized incident electromagnetic waves. **e** and **f** show the measured transmission spectra of SSR metasurfaces with x- and y-polarized incident electromagnetic waves, respectively

strength increases when the meta-molecules radiate the incident THz waves via electrical dipole resonances. Therefore, the Fano resonance induced by x-polarized light reaches its maximum strength at the MIDDLE-state when $\Delta s = 15\ \mu\text{m}$.

On the other hand, the electrical dipole resonances are induced simultaneously on both SRR and the slab structures under y-polarized incidence. The quadrupole resonance modes induce the phase difference between the resonances on the SRR and slab structures. The electrical dipole resonances on SRR and slab structures are in phase when the meta-molecules are tuned to the OPEN-state and the CLOSE-state, which results in strong dipole resonances. Therefore, the Fano resonance under y-polarized incidence reaches its maximum strength at the OPEN-state and the CLOSE-state, which is opposite to that under the x-polarized incidence.

4.5 Summary

In recent years, active photonic devices have been facing tremendous challenges from the ever-increasing demands of the market in terms of switching speed, tuning range, flexibilities, etc. The reconfigurable meta-molecules offer a new method to control the electrical and magnetic resonances in subwavelength resolution. Many extraordinary electromagnetic properties, e.g., EIT-like effect and Fano resonance, have now been realized and controlled by the structural reconfigurable metasurfaces. More importantly, meta-devices with promising functionalities can be realized with the development of MEMS technologies.

References

1. Jacobs M, Lopez-Garcia M, Phrathep OP, Lawson T, Oulton R, Whitney HM (2016) Photonic multilayer structure of Begonia chloroplasts enhances photosynthetic efficiency. Nature Plants 2(11):16162. https://doi.org/10.1038/nplants.2016.162
2. Koonath P, Jalali B (2007) Multilayer 3-d photonics in silicon. Opt Express 15(20):12686–12691. https://doi.org/10.1364/OE.15.012686
3. Lugo JE, Lopez HA, Chan S, Fauchet PM (2002) Porous silicon multilayer structures: a photonic band gap analysis. J Appl Phys 91(8):4966–4972. https://doi.org/10.1063/1.1461898
4. Nascimento EM, Zanetti FM, Lyra ML, de Oliveira IN (2010) Tunable reflectance spectra of multilayered cholesteric photonic structures with anisotropic defect layers. Phys Rev E 81(3):031713. https://doi.org/10.1103/PhysRevE.81.031713
5. Whittaker DM, Culshaw IS (1999) Scattering-matrix treatment of patterned multilayer photonic structures. Phys Rev B 60(4):2610–2618. https://doi.org/10.1103/PhysRevB.60.2610
6. Cheng Y, Sun M (2021) Unified treatments for localized surface plasmon resonance and propagating surface plasmon polariton based on resonance modes in metal nanowire. Optics Commun 499:127277. https://doi.org/10.1016/j.optcom.2021.127277
7. Liu W-C, Tsai DP (2002) Optical tunneling effect of surface plasmon polaritons and localized surface plasmon resonance. Phys Rev B 65(15):155423. https://doi.org/10.1103/PhysRevB.65.155423
8. Luo X, Ishihara T (2004) Subwavelength photolithography based on surface-plasmon polariton resonance. Opt Express 12(14):3055–3065. https://doi.org/10.1364/OPEX.12.003055
9. Mayer KM, Hafner JH (2011) Localized surface plasmon resonance sensors. Chem Rev 111(6):3828–3857. https://doi.org/10.1021/cr100313v
10. Mojarad NM, Agio M (2009) Tailoring the excitation of localized surface plasmon-polariton resonances by focusing radially-polarized beams. Opt Express 17(1):117–122. https://doi.org/10.1364/OE.17.000117

11. Murray WA, Astilean S, Barnes WL (2004) Transition from localized surface plasmon resonance to extended surface plasmon-polariton as metallic nanoparticles merge to form a periodic hole array. Phys Rev B 69(16):165407. https://doi.org/10.1103/PhysRevB.69.165407
12. Willets KA, Van Duyne RP (2007) Localized surface plasmon resonance spectroscopy and sensing. Annu Rev Phys Chem 58(1):267–297. https://doi.org/10.1146/annurev.physchem.58.032806.104607
13. Koshelev K, Lepeshov S, Liu M, Bogdanov A, Kivshar Y (2018) Asymmetric metasurfaces with high-Q resonances governed by bound states in the continuum. Phys Rev Lett 121(19):193903. https://doi.org/10.1103/PhysRevLett.121.193903
14. Meinzer N, Barnes WL, Hooper IR (2014) Plasmonic meta-atoms and metasurfaces. Nat Photonics 8(12):889–898. https://doi.org/10.1038/nphoton.2014.247
15. Nouman MT, Hwang JH, Jang J-H (2016) Ultrathin terahertz quarter-wave plate based on split ring resonator and wire grating hybrid metasurface. Sci Rep 6(1):39062. https://doi.org/10.1038/srep39062
16. Patel SK, Sorathiya V, Nguyen TK, Dhasarathan V (2020) Numerical investigation of tunable metasurface of graphene split-ring resonator for terahertz frequency with reflection controlling property. Physica E 118:113910. https://doi.org/10.1016/j.physe.2019.113910
17. Pendry JB (2000) Negative refraction makes a perfect lens. Phys Rev Lett 85(18):3966–3969. https://doi.org/10.1103/PhysRevLett.85.3966
18. Pendry JB, Aubry A, Smith DR, Maier SA (2012) Transformation optics and subwavelength control of light. Science 337(6094):549–552. https://doi.org/10.1126/science.1220600
19. Smith DR, Pendry JB, Wiltshire MCK (2004) Metamaterials and negative refractive index. Science 305(5685):788–792. https://doi.org/10.1126/science.1096796
20. Boller KJ, Imamoğlu A, Harris SE (1991) Observation of electromagnetically induced transparency. Phys Rev Lett 66(20):2593–2596. https://doi.org/10.1103/PhysRevLett.66.2593
21. Fleischhauer M, Imamoglu A, Marangos JP (2005) Electromagnetically induced transparency: optics in coherent media. Rev Mod Phys 77(2):633–673. https://doi.org/10.1103/RevModPhys.77.633
22. Marangos JP (1998) Electromagnetically induced transparency. J Mod Opt 45(3):471–503. https://doi.org/10.1080/09500349808231909
23. Röhlsberger R, Wille H-C, Schlage K, Sahoo B (2012) Electromagnetically induced transparency with resonant nuclei in a cavity. Nature 482(7384):199–203. https://doi.org/10.1038/nature10741

Chapter 5
Tunable Optical Anisotropic Metasurfaces with Dynamic Control of In-Plane Symmetry

5.1 Optical Anisotropy

Light is an electromagnetic wave with transverse vibrations of electrical and magnetic fields, the directions of which define its polarization state [1–3]. Light is sometimes described as unpolarized from a macroscopic point of view, which only means that it contains a mixture of light polarized in different angles without a net polarization. The polarization state of the light may affect the light-matter interaction process for certain materials, which is similar to the interaction between a radio wave and a polarization-selective antenna [4–6]. For example, the radio waves can be collected by exciting the electronic resonances within an antenna which converts the electromagnetic waves to electrical signals. However, the antenna is polarization-selective when the electronic resonances can only be excited by the electrical field in a specific direction, thus limiting the receivable radio waves to one polarization state.

Similarly, the electrons in a crystal can be driven by external electromagnetic fields with different polarization states. In some crystals, the moving of the electrons is dependent on the directions of the electromagnetic waves, i.e., their polarization states, due to the crystal structures, which result in the difference in the propagating speed between two orthogonal polarization states [7, 8]. On the other hand, the refractive index of a material is defined as the velocity of the light propagating in a vacuum divided by that of the light propagating in the material. As a result, the refractive index of a crystal is dependent on the polarization state when the electrons selectively delay the propagating speed of the light according to its polarization, which is defined as birefringence. The birefringent crystals have two refractive indices along with two directions. Light propagating along one direction has higher refractive indices than the other, which are defined as n_e and n_o, respectively. Here "e" and "o" stand for extraordinary and ordinary, respectively. Rasmus Bartholin first described the birefringence effect in 1669, who found the double refraction in Iceland spar (calcite) [9]. The relation between the polarization of the light and the birefringence effect was discovered and explained by Augustin-Jean Fresnel in the nineteenth century [10, 11]. As a characteristic of a material, birefringence is defined as the difference

W. Zhu and A.-Q. Liu, *Metasurfaces: Towards Tunable and Reconfigurable Meta-devices*, Microfluidics and Nanophotonics: Science and Engineering 1,
https://doi.org/10.1007/978-981-19-6925-6_5

between the maximum and minimum refractive index values, which is a measure of optical anisotropy. The birefringence is mainly due to the asymmetry of the crystal structures of natural materials, which is inherently weak. The most highly birefringent materials in the optical frequency region are rutile and calcite, which have a birefringence of 0.287 and 0.190, respectively. Other birefringent materials, such as quartz and corundum, have a birefringence of approximately 0.01 [12–16].

Optical anisotropy, especially controllable optical anisotropy, has been widely applied to many devices. For example, light modulators can be realized by incident polarized light on electrically controlled optical anisotropic materials, e.g., liquid crystals, where the electrically induced birefringence tunes the transmission of the polarized light. This method can also be applied to unpolarized light by cascading the tunable optical anisotropic material with a polarizer, which has been widely used in liquid–crystal displays [17, 18]. Optical anisotropy is typically due to the interactions between the light and the atoms, which delay the light propagating speed by the absorption and re-radiation processes. This process defines the refraction indices of the crystal materials, which are highly dependent on the spatial symmetry of the lattice structure. With the development of nanofabrication technologies, space-variant permittivities and permeabilities of artificial materials have been demonstrated by using subwavelength structures composed of isotropic materials. With rational designed micro/nano structures, the effective refractive indices of the artificial materials can be derived by using the electrostatic field approximation method, which shows similar properties as anisotropic crystals, e.g., birefringences. Later, metamaterials and metasurfaces are also demonstrated to have optical anisotropy with meta-molecules or lattice structures designed to have resonance modes dependent on the polarization states of the incidence [19–21]. These anisotropic artificial materials have optical properties dependent on the incident polarization states, which are called "form birefringence" or "artificial optical anisotropy". The artificial optical anisotropy can be predesigned by varying the geometries of the meta-molecules or their spatial distributions, which offer significantly improved design flexibilities for devices based on optical anisotropy.

5.2 Maltesecross Metamaterial

Metasurfaces with tunable and arbitrary optical anisotropy have vast applications such as wave plates, birefringent filters, optical isolators, etc. However, tunable metasurfaces with an extensive tuning range of optical anisotropy are challenging to be realized due to the difficulty in changing the symmetry of the metamaterials microstructures once fabricated.

In this section, a micromachined structural reconfigurable metasurface is tuned between positive anisotropy, negative anisotropy, and isotropy states. The transmission spectrum of extraordinary (*e*-polarized) incidence can be tuned while maintaining that of the ordinary (*o*-polarized) incidence by continuously shifting one trapezoid of the cross-shaped meta-molecules. In experiments, it measures a Fano resonance frequency shift of 25.9% and 20.4% at low frequency and high-frequency regions, respectively, for *e*-polarized incidence and only 0.8% for *o*-polarized incidence.

The tunable metasurface consists of cross-shaped meta-molecules with two dimensional (2D) square-lattice array (Fig. 5.1a). The Maltese-cross meta-molecule has four trapezoid metal strips with a height h of 11 μm, the longer parallel side L_1 of 4 μm, and the shorter parallel side L_2 of 1 μm. The square lattice has a period P of 28 μm. One trapezoid metal strip in each meta-molecule is supported by movable silicon frames (green) and driven by two micromachined comb-drive actuators. The rest of the meta-molecules locate on the silicon substrate (blue) anchored on the SOI wafer. With mechanical actuation, the movable trapezoid can move along the y-direction with a displacement up to 5 μm. The normal incidence has a wavevector k perpendicular to the metasurface, while the polarization states are defined by the orientation directions of the electrical fields. The extraordinary polarization (*e*-polarization) and ordinary polarization (*o*-polarization) refer to the polarization states of the incidence when its electric field is along (y-direction) and perpendicular to (x-direction) the actuation direction of the movable trapezoid, respectively. Insert shows the cross-view of the meta-molecule. Figure 5.1b, c, d show the schematics of the meta-molecules when the displacements of movable trapezoid S are 0 μm, 2.5 μm, and 5 μm, respectively.

The transmission spectra of the tunable anisotropic metasurface under different displacements are shown in Fig. 5.2 for both *e*-polarized (red-solid line) and *o*-polarized (blue-dotted line) incident THz waves. Figure 5.2a shows that the transmission spectra of both e and *o*-polarized incidence are identical, which is due to the four-fold rotational symmetry of the Maltese-cross meta-molecule at the initial state. The Fano resonance dips are at 2.45 THz for *e*-polarized and *o*-polarized incidence. However, two Fano resonance dips are observed under *e*-polarized incidence when the meta-molecule is changed to the middle state (Fig. 5.2b) and the final state (Fig. 5.2c). At the middle state, the Fano resonance dips for *e*-polarized light are 2.70 THz in the low-frequency region and 4.48 THz in the high-frequency region. The Fano resonance dip shifts to 2.00 THz in the low-frequency region and 4.21 THz in the high-frequency region when the meta-molecule is at the final state.

The contour maps of the surface current at the Fano resonance dip frequencies are calculated to explain the origin of the resonance modes further when the shift distances S are 0 μm, 2.5 μm, and 5 μm, respectively. The color represents the surface current density, while the arrows show the surface current flux direction. The first and second columns show the surface current distribution at the resonance frequency under *e*-polarization incidence, while the third column shows the surface current distribution under *o*-polarization incidence. The last column shows the coupled oscillators corresponding to each state. When $S = 0$ μm, the incidences excite the two trapezoid metal strips oriented along the incident electric field, as shown in Fig. 5.2a. Each trapezoid is represented by one classic oscillator in the coupled oscillators. The resonance is excited along the direction of the electrical fields. The excited resonance modes couple to two dark modes out of phase and result in zero net effect. Corresponding to the three-level system, only the transitions between level 1 (L_1) and level

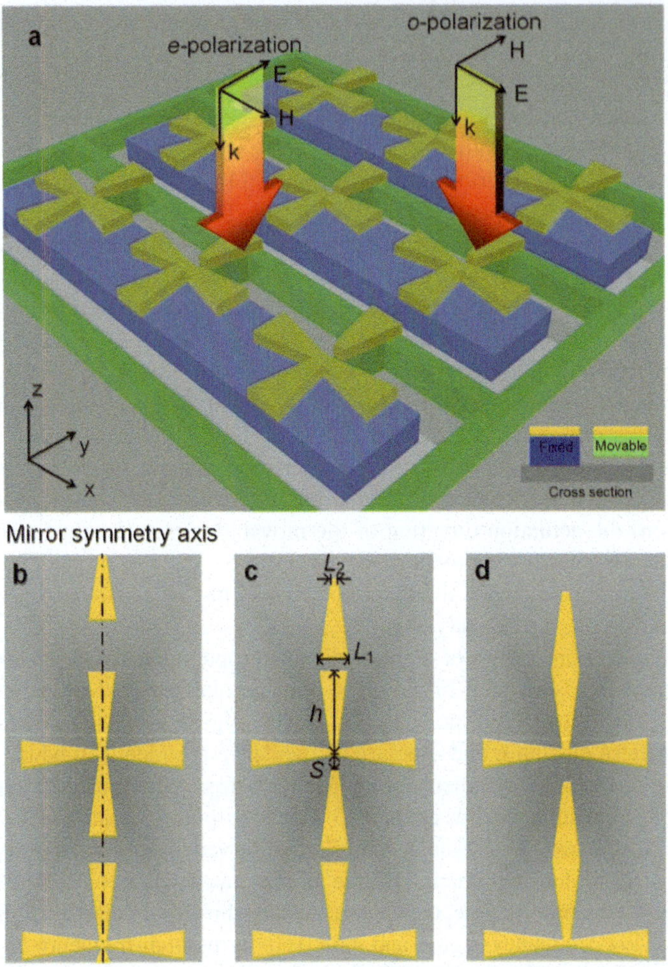

Fig. 5.1 Schematics of the tunable anisotropic metasurface with Maltesecross meta-molecules **a** The overview of the metasurface. The cross-shaped meta-molecules consist of four trapezoid metal stripes, one of which locates on the movable frames driven by the comb-drive actuators (green). The rest of the meta-molecules locate on the silicon substrates (blue) fixed on the SOI wafer. Insert shows the cross-view of the meta-molecules. **b**, **c** and **d** show the meta-molecules when the tunable anisotropic metasurface is at the initial state $S = 0$ μm, the middle state $S = 2.5$ μm, and the final state $S = 5$ μm, respectively. Reproduced with permission [22] copyright from Springer Nature

2 (L_2) are excited. The Fano resonance dips only appear in the low-frequency region since the two excited trapezoids are connected and form a large resonator. When $S = 2.5$ μm, the two trapezoid strips along the *y*-direction are disconnected, resulting in a blue shift of the Fano resonance dip (Fig. 5.2b).

Furthermore, *e*-polarized incidence induces a dark mode resonance within the trapezoid along the *x*-direction, corresponding to the transition between level 1 and

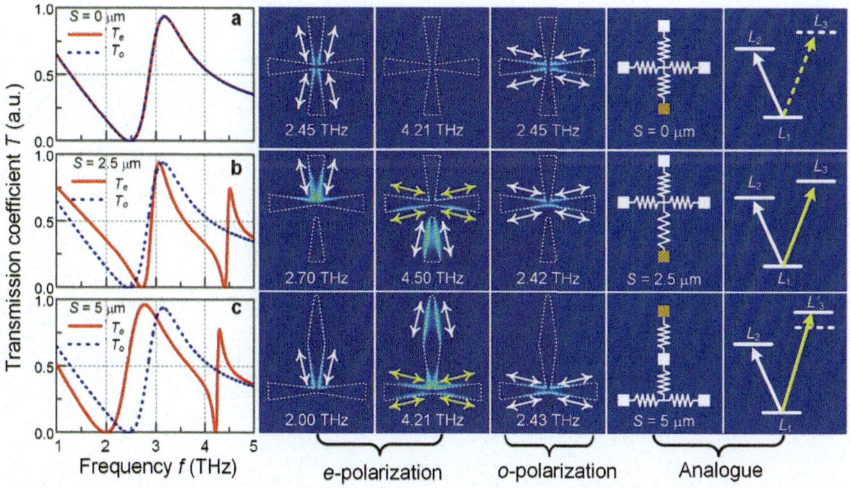

Fig. 5.2 Transmission spectra of the tunable anisotropic metasurface for *e*-polarized (red-solid line) and *o*-polarized (blue-dotted line) incidences when the shift distance is **a** 0 μm, **b** 2.5 μm, and **c** 5 μm. The contour maps show surface currents under the excitation of different incidence frequencies. The color and the arrows represent the intensity and the direction of the surface current. The white and yellow shades represent the surface currents excited directly by the incidence electric field and indirectly by the coupling between the trapezoids, respectively. The first and second columns show the surface current distribution at the resonance frequency under *e*-polarization incidence, while the third column shows the surface current distribution under *o*-polarization incidence. The last column shows the coupled oscillators corresponding to each state. The fourth column shows the three-level system corresponding to the resonance mode excited by the electric field (white) and the coupling between the trapezoids (yellow), respectively. The interference between the two modes results in the Fano resonance in the high-frequency region

level 3 in the three-level system. The second Fano resonance arises at the high-frequency region (4.48 THz) due to the interference between dark and bright modes. When $S = 5$ μm, the movable trapezoid strip attached to the rest of the meta-molecule from the backside formed a large metal strip, resulting in a red-shift of the Fano resonance at the low-frequency region for *e*-polarized incidence as shown in Fig. 5.3c. Similarly, the resonance frequencies of the bright modes are also red-shifted in the high-frequency area. The displacement of the movable trapezoid has minor effects on the resonance induced by *o*-polarized incidence, resulting in similar Fano resonances frequency along the *x*-direction.

Two frequency samples, which are 2.9 THz at the low-frequency region and 4.5 THz at the high-frequency region, are chosen to analyze the evolution of the optical anisotropy for the tunable metamaterials when S is increased from 0 to 5 μm. The linear dichroism $\Delta = A_e - A_o$ is defined as the difference in the extinction ratio between the *e*-polarized ($A_e = 1 - T_e - R_e$) and *o*-polarized ($A_o = 1 - T_o - R_o$) incidence where T and R stand for the transmission and reflection coefficient, respectively. The linear dichroism is maintained below 1.5% during the shifting of the movable trapezoid, which is shown in Fig. 5.3a. The optical anisotropy is quantified

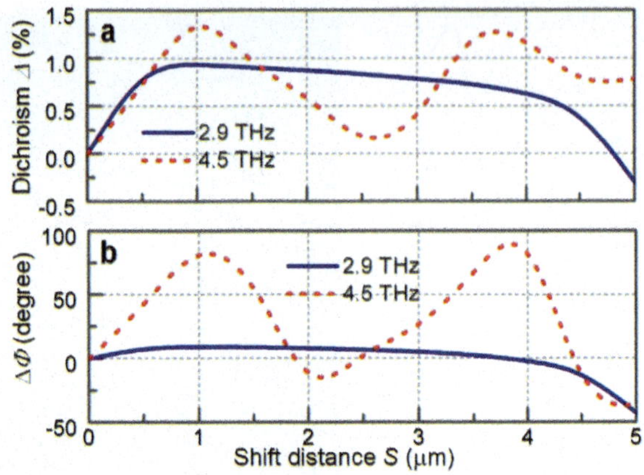

Fig. 5.3 **a** The linear dichroism and **b** difference of phase change $\Delta\Phi$ between *e*-polarization and *o*-polarization ray ($\Delta\Phi = \Phi_e - \Phi_o$) as the function of the shift distance *S*. The blue solid and red dotted lines represent the incident frequency of 2.9 THz and 4.5 THz, respectively

using the difference of the phase shift ($\Delta\Phi = \Phi_e - \Phi_o$) between the *e*-polarized (Φ_e) and *o*-polarized (Φ_o) incidence. $\Delta\Phi$ is tuned between 9° to − 42° and 116° to − 37° at 2.9 THz and 4.5 THz, respectively, as shown in Fig. 5.3b. The evolution of $\Delta\Phi$ shows the tuning of the tunable metamaterials between negative anisotropy, positive anisotropy, and isotropy states at both 2.9 and 4.5 THz.

The structures of the tunable metamaterial are fabricated on a silicon-on-insulator (SOI) wafer using the deep reactive ion etching (DRIE) processes. Figure 5.4a shows the overview SEM graph of the tunable anisotropic metasurface. Two identical micromachined comb-drive actuators driven by electrostatic forces are placed on both sides of a 400 × 400 meta-molecule array with a footprint of approximately 1 cm^2. Each actuator provides bidirectional in-plane translation (along *x*-direction). Figure 5.4b shows a closed-view SEM graph of the meta-molecules formed by patterning a 0.5-μm thick aluminum layer on the top of the SOI wafer. The movable trapezoids are patterned on the silicon frame, consisting of many narrow beams (3-μm width) with tips. The fixed parts of the meta-molecules are patterned on the isolated anchors. The silicon frame is fully released and becomes freely movable while the anchor remains fixed on the substrate by controlling the release time.

The transmission spectra at different displacements are measured using a Bruker Vertex 80v Fourier transform infrared (FTIR) to characterize the lattice displacement effects of the fabricated tunable anisotropic metasurface. Figure 5.5 shows the measured transmission spectra for *e*-polarized (first column) and *o*-polarized (second column) incidence. The inserts show the schematic of the meta-molecules. When $S = 0$ μm, the Fano resonance dip is measured to be 2.42 THz and 2.41 THz for *e*-polarized and *o*-polarized incidence, respectively (Fig. 5.5a, b). When $S = 2.5$ μm, the measured Fano resonance dip is 2.65 THz (low-frequency region) and 4.35 THz (high-frequency region) under *e*-polarized incidence (Fig. 5.5c). When $S = 5$ μm, the Fano resonance dips under *e*-polarized incidence are 1.94 THz in the low-frequency region and 4.17 THz in the high-frequency region, both of which are red-shifted compared with the Fano resonances when $S = 2.5$ μm as shown in Fig. 5.5e. The

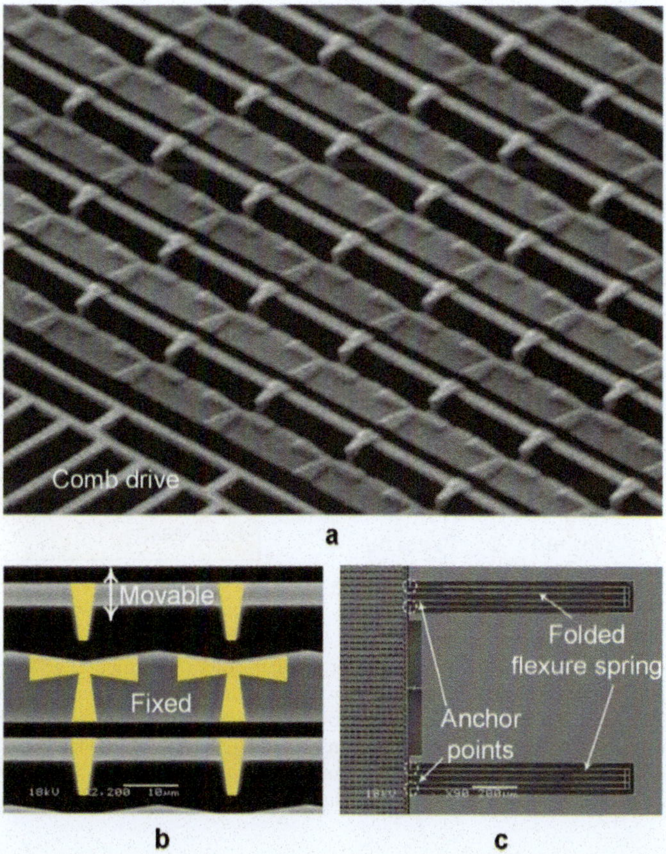

Fig. 5.4 Scanning electron microscopy (SEM) graph of the micromachined tunable metamaterial. **a** Overview of the meta-molecules array and comb drive. **b** Zoomed-in view of the meta-molecules (The metal part is highlighted with false color). **c** The zoomed-in view of the comb-drive actuator

variation of Fano resonance dip is within 0.02 THz (0.08% of resonance frequency 2.39 THz) for *o*-polarized incidence.

Contour maps of transmission coefficients for *e*-polarized (Fig. 5.6a) and *o*-polarized (Fig. 5.6b) incidences are shown to illustrate further the variation of the optical anisotropy due to the shift distance of movable trapezoid metal strip. The bright region shows the high transmission region, while the dark area shows the transmission dip. The measured Fano resonance dips are marked by white circles that agree with the simulation results. The Fano resonance dip frequency is firstly blue-shifted and then red-shifted for *e*-polarized incidence when the shift distances of a movable trapezoid increase from 0 to 5 μm. The tuning ranges of the Fano resonance dip frequency are 0.75 THz and 0.92 THz, respectively.

In conclusion, micromachined reconfigurable metamaterials which can be tuned between positive anisotropy ($\Delta\Phi = 116°$), negative anisotropy ($\Delta\Phi = -37°$),

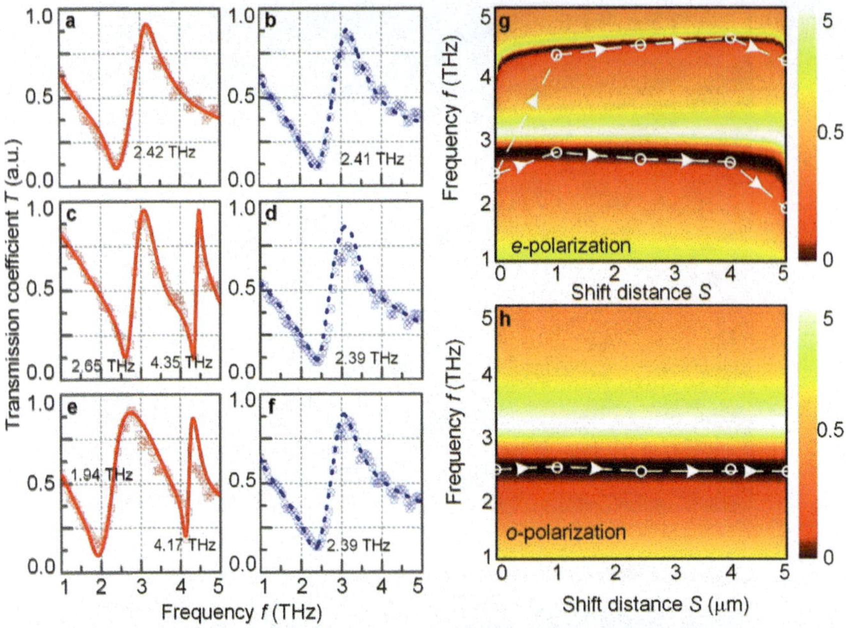

Fig. 5.5 **a–f** Measured transmission spectra for *e*-polarization (first column) and *o*-polarization (second column) incidences. The initial state S = 0 μm, middle state S = 2.5 μm, and final state S = 5 μm are shown in the first, second, and last rows of transmission spectra, respectively. Contour map of transmission coefficient for *e*-polarization (**g**) and *o*-polarization (**h**) incidences. The bright and dark regions represent the high transmission and the transmission dip, respectively. White circles mark the measured Fano resonance dips to compare with the simulation results

and isotropy states ($\Delta\Phi = 0$) is, experimentally demonstrated. The Fano resonance frequency shifts are measured to be 25.9% and 20.4% for *e*-polarized incidence and only 0.8% for *o*-polarized incidence. The tunable metasurface with an extensive tuning range of optical anisotropy has vast applications such as wave plates, birefringent filters, light modulators, etc.

5.3 Lattice Constant Variation of THz Metamaterials

Metasurfaces can be designed to have uniform optical properties, i.e., permeabilities and permeabilities, by using identical meta-molecules with periodic lattice structures [23]. Similar to natural materials, extraordinary optical properties, e.g., birefringences [24], optical chirality [25, 26], dichroism [27, 28], etc., can be obtained by tailoring the symmetries of the metasurface lattice structures. In this section, an example of MEMS metasurfaces is discussed, which has tunable anisotropy induced by the variation of the lattice structures.

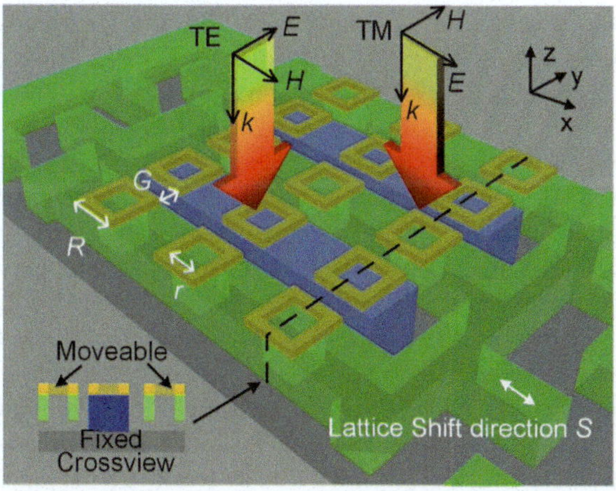

Fig. 5.6 The schematic of the tunable anisotropic metasurface with lattice variation. The movable group of meta-molecules (golden) is patterned on the silicon frames (green), which can be actuated along the x-direction. The fixed group of meta-molecules is patterned on the fixed substrates (blue), which are anchored on the SOI wafer (gray). The insert shows a cross-view of the metasurface. Reproduced with permission [29] copyright from AIP Publishing

The tunable anisotropic metasurface consists of micro-ring-shaped meta-molecules with a 2D rectangular lattice array, which is shown in Fig. 5.6. The inner and outer radius of the micro-ring-shaped meta-molecule is designed to be 12 μm and 18 μm, respectively, which are designed to work in the THz region. Similar to the MEMS metasurfaces discussed in previous chapters, the meta-molecules are divided into two groups by every other line along the y-direction. One group of meta-molecules is located on the movable silicon frames, which are represented by green color in Fig. 5.6. These meta-molecules can be actuated along the x-direction by using comb-drive actuators. The other group of meta-molecules is located on silicon substrates (blue), which are fixed on the SOI wafer as highlighted by the gray color. The cross-view of meta-molecules is shown in the inserted graph in Fig. 5.6. In the initial state, the micro-ring-shaped meta-molecules form a rectangular lattice array with translation symmetry along both x and y-directions but with different lattice constants. The period along the x-direction P_x is 56 μm, while the period along the y-direction P_y is 28 μm. Here the period along the x-direction is deliberately designed to be twice the length of the period along the y-direction so that the lattice structure can be tuned from asymmetry state to symmetry state by mechanical actuation of the movable meta-molecules.

The incident THz wave is along the z-direction, which is perpendicular to the metasurface. Here, the incident polarization states are defined by the orientation direction of the electrical fields, where the TE and the TM polarization states refer to the incidence with electrical field along y and x-directions, respectively. The meta-molecules within the movable group can be actuated simultaneously along the x-direction, which changes their relative positions to the meta-molecules within the fixed group.

The actuation displacement of the movable group of meta-molecules is denoted by *S*, which has an effective length of 28 μm due to the translation symmetry. The displacement of the movable group changes the coupling conditions between the movable and fixed meta-molecules, which results in the variations of the surface currents induced by the incident THz waves, as shown in Fig. 5.7a, b. Due to the invariance of the meta-molecules, the surface currents depend solely on the lattice structure of the metasurface. Figure 5.7a, b show the contour maps of meta-molecules when the lattice structure is at the initial state (polarization-dependent) and polarization-independent state, respectively. Here the polarization-independent state is defined as the lattice structure with displacement $S = 28$ μm when the lattice has four-fold rotational symmetry and responses to TE and TM polarized incidence identically. The contour maps of surface currents are simulated under TE polarized incident with a frequency of 3.11 THz. The inserts show the directions of surface currents, which are represented by arrows. Figure 5.7a shows the surface current is feeble at the initial state, while Fig. 5.7b shows a strong surface current at the polarization-independent state. Since the meta-molecules are designed to be closed rings without gaps, the LC resonances are not applicable within this frequency region with the normal incidence. The surface currents are induced purely by the dipole resonances along the electrical field direction. As a result, the meta-molecule are excited by the incident electrical field only.

Figure 5.7c, d show the transmission spectra of the tunable anisotropic metasurface with different lattice displacements under TE and TM-polarized incidence, respectively. Due to the translation symmetry, the lattice structures are the same when the difference between two lattice displacements is 28N μm, where N is an arbitrary integer number. The lattice displacement is chosen between 0 and 56 μm in the simulation and 0–28 μm in the experiment. The transmission spectra of the tunable anisotropic metasurface when $S = 28$ μm, $S = 14$ μm, and $S = 0$ μm are represented by the red-solid, green-dashed, and blue-dotted lines, respectively. Figure 5.7c shows that the transmission dips are blue-shifted from 2.47 to 3.11 THz when the displacement *S* increases from 0 to 28 μm. On the other hand, the transmission dips are red-shifted from 3.53 to 3.11 THz when the displacement *S* increases from 0 to 28 μm for TM-polarized incidence. The period of the lattice changes dramatically along the y-direction since the lattice shift is along the x-direction. As a result, TE-polarized incidence is more sensitive to the lattice variation than the TM-polarized incidence. Another resonance mode appears during the actuation process when S is changed from 0 to 28 μm. This resonance mode is proven by the transmission dip at 3.32 THz when S = 14 μm. It should be pointed out that the transmission spectra are identical for both TE and TM-polarized incident THz waves when S = 28 μm, which is the polarization-independent state with four-fold rotational symmetry of the lattice structure.

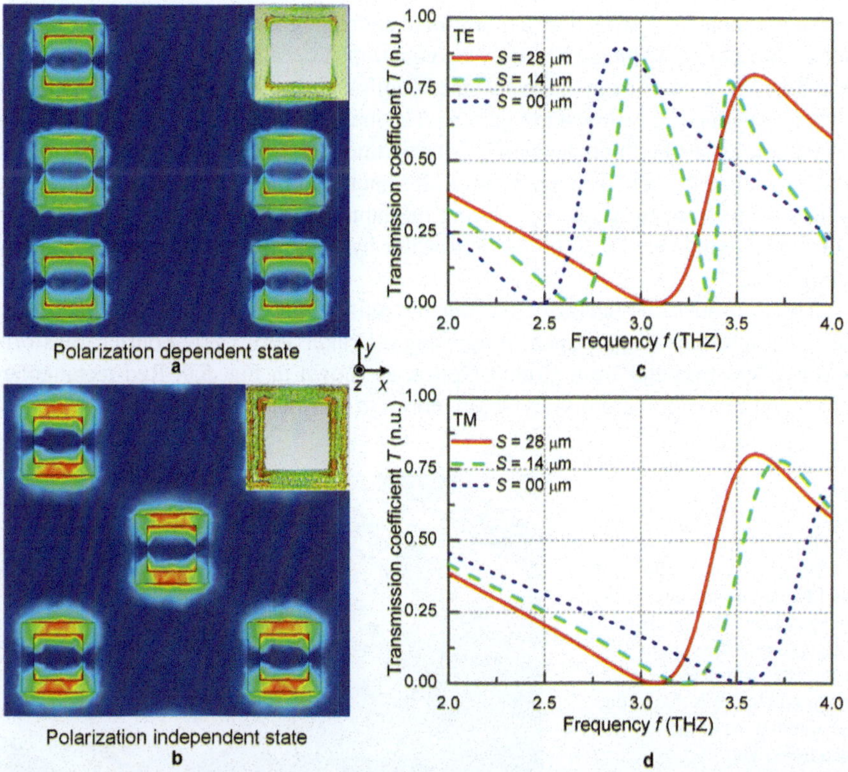

Fig. 5.7 Numerical simulation results of the tunable anisotropic metasurface. The contour maps of the surface current of TE-polarized incidence **a** $S = 0$ μm and **b** $S = 28$ μm. The incident frequency is chosen to be 3.11 THz. The strength of the surface current is represented by the color, while the inserts show the instantaneous directions of the surface currents. **c** and **d** show the transmission spectra under different lattice shifts with TE and TM-polarized incidences, respectively. The red-solid, green-dashed, and blue-dotted lines represent the transmission spectra when S = 28 μm, S = 14 μm, and S = 0 μm, respectively

The tunable anisotropic metasurface is fabricated by using the DRIE processes on an SOI wafer, which is shown in Fig. 5.8. The overview SEM graph of the metasurface is demonstrated in Fig. 5.8a. The aluminum meta-molecules are highlighted with false yellow color. The supporting structures for the movable group consist of silicon frames of 3 μm in width, which is the same as the comb-drive structures as shown at the bottom of the graph. The meta-molecules of the fixed group are anchored on silicon substrates of 28 μm in width, which cannot be removed from the SOI wafer during the release process. Here, the silicon substrates of the fixed meta-molecules are isolated from each other, which are brighter than the silicon frame in the SEM graphs due to the accumulations of the electrons. The movable meta-molecules are mechanically actuated by two identical comb-drive actuators, which locate on both sides of the metasurface along the x-direction. The metasurface consists of a 400 ×

200 meta-molecules array with a footprint of approximately 1 cm^2. The displacement S and the actuation voltage V satisfy the equation $\Delta x = AV^2$ where $A = 0.05$ μm/V^2 is the actuation coefficient of the designed comb-drive actuator. A zoomed-in view of the meta-molecules is shown in Fig. 5.8b. The meta-molecule in the middle is formed by patterning an aluminum layer of 0.5-μm in thickness on the movable frame, which is sandwiched by two fixed meta-molecules anchored on the silicon substrates. The silicon frames are fully released during the fabrication processes while the silicon substrates remain anchored on the SOI wafer by precisely controlling the releasing time.

The transmission spectra of the tunable anisotropic metasurface are measured at different lattice displacement S by using a Bruker Vertex 80v Fourier transform infrared spectrometer from Bruker Optics, as shown in Fig. 5.9. Red-rectangular-cross and blue-circular-cross symbols represent the measured transmission spectra

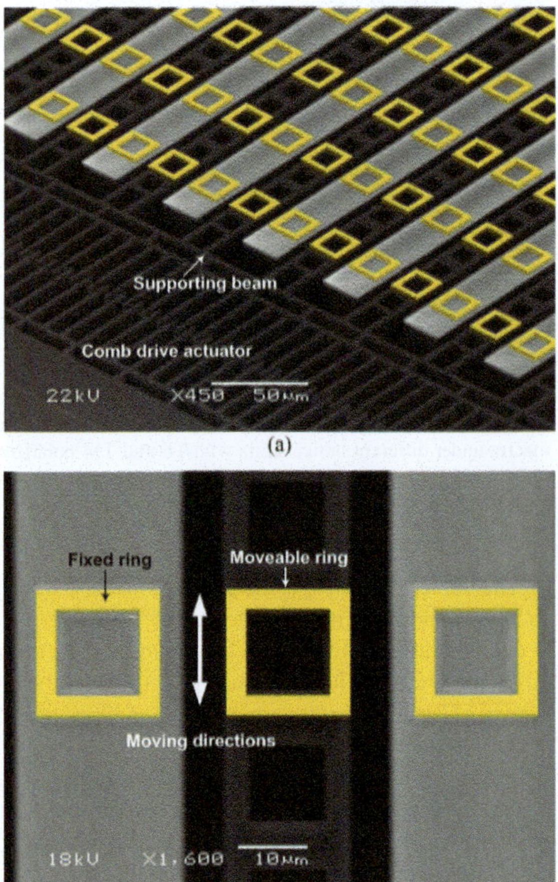

Fig. 5.8 Scanning electron microscopy (SEM) graph of the tunable anisotropic metasurface. The **a** overview and **b** zoomed-in view of the metasurface. The enhanced brightness of the silicon substrates with fixed meta-molecules is due to the accumulations of the electrons. The meta-molecules are highlighted with false yellow color. Reproduced with permission [29] copyright from AIP Publishing

of TE and TM-polarized incident THz waves, respectively. For comparison, the simulated transmission spectra of TE and TM-polarized incident THz waves are represented by red-solid lines and blue-dashed lines, respectively. Figure 5.9a–c show the transmission spectra when $S = 28\ \mu\text{m}$, $S = 14\ \mu\text{m}$, and $S = 0\ \mu\text{m}$, respectively. The transmission spectra of TE and TM-polarized incidence are overlapped with each other when the lattice displacement $S = 28\ \mu\text{m}$. The metasurface is tuned to the isotropic (polarization-independent) state when the lattice structure has four-fold rotational symmetry. The transmission spectra of TE and TM-polarized incidence are quite different from each other at the initial state when $S = 0\ \mu\text{m}$.

The slight mismatch between transmission spectra of TE and TM-polarized incidence is due to the asymmetry of the supporting structures of the movable and fixed meta-molecules, i.e., the silicon substrates and frames. The transmission dip is shifted to a higher frequency when the polarization state of the incident THz wave changes from TE to TM. The resonance dips depend on the lattice structures, especially the lattice constant along the direction of incident electric fields since the meta-molecules are only excited by the electrical fields. The variations of the lattice constant can explain the shifts of the resonance dips due to the lattice displacement. For example, the lattice constant is 56 μm at the initial state (S = 0 μm) along the y-direction, the direction of the electrical fields for TE-polarized incidence. The effective distance between two adjacent meta-molecules increases as the lattice displacement increases. On the other hand, the resonance dips are blue-shifted from 2.48 to 2.62 THz when the lattice displacement increases. It can be assumed that the electrically induced resonance frequency is proportional to the lattice constant along the direction of the incident electrical fields. The trivial variation of the transmission dips also verifies this assumption with TM-polarized incidence since the lattice constant along the x-direction remains 28 μm during the actuation process. It should be pointed out that another resonance dip arises during the tuning process for TE-polarized incidence, which is due to the dramatic change of the lattice along the y-direction. The experimental results agree well with the simulation results.

The evolution of the transmission spectra during the tuning process of the anisotropic metasurface can be illustrated by the contour maps, as shown in Fig. 5.9, where the color represents the transmission coefficients. Figure 5.9d, e show the contour map of TE-polarized and TM-polarized incidence with the lattice displacement ranging from 0 to 56 μm. Both Fig. 5.9d, e are symmetric with an axis at $S = 28\ \mu\text{m}$ (half of the lattice constant along the actuation direction) due to the translation symmetry of the lattice. Figure 5.9f shows the combined contour map of Fig. 5.9d, f, which match each other perfectly at $S = 28\ \mu\text{m}$. As a result, the anisotropic metasurface is tuned from a polarization-dependent state to a polarization-independent state (isotropic state).

In conclusion, the THz tunable anisotropic metasurface is designed and experimentally demonstrated using MEMS metasurfaces with reconfigurable lattice structures. The transmission dips of TE-polarized THz waves are tuned from 2.48 to 3.12 THz, with the lattice displacement ranging from 0 to 28 μm. The transmission dips of TM-polarized THz waves are shifted from 3.55 to 3.12 THz with a tuning range smaller than that of the TE-polarized incidence due to the lattice constant remaining

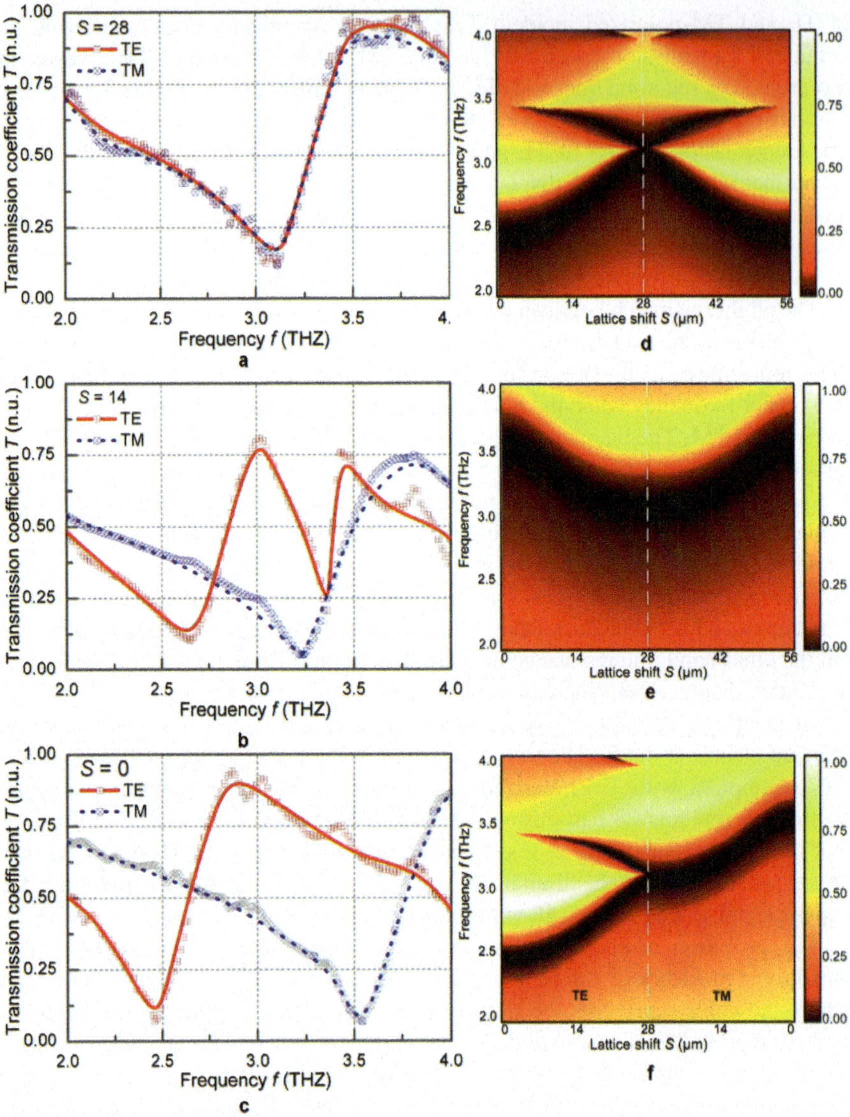

Fig. 5.9 Experimental and simulation results of the transmission spectra for TE-polarized (red) and TM-polarized (blue) incidences, respectively. **a**, **b** and **c** show the transmission spectra when lattice displacements are 28 μm, 14 μm, and 0 μm, respectively. Contour map of transmission coefficient for TE (**d**) and TM (**e**) polarized incidences. The bright region shows the high transmission region, while the dark part shows the transmission dip. The combined contour map of (d) and (e) is shown in (**f**)

in the same direction of incident electrical fields during the tuning process for TM-polarized incidence. This work proves that the tunable anisotropy of metasurfaces can be realized by changing their lattice symmetry with MEMS actuators.

5.4 Summary

The anisotropic microstructures of atoms or molecules often result in optical anisotropy in nature materials [30]. For example, the uniaxial anisotropic structure is such that it has an axis of symmetry with no equivalent axis in the perpendicular plane. Although nature materials have optical anisotropy ranging from negative to positive values, it is challenging to find materials with arbitrary optical anisotropy in nature. However, the microscopic structures of artificial materials, such as metamaterials, can be rationally designed to achieve unique optical anisotropy, which is not ready in nature [31]. For many applications, such as display, light modulation, tunable birefringent filter, etc., tunable optical anisotropy is almost necessary [32]. For example, metamaterials with tunable optical anisotropy can alter the effective optical path of light with different polarization, which is traditionally achieved by changing the thickness of the birefringent materials used in the birefringent filters or waveplates.

However, most tunable metamaterials are tuned by refractive index variation of the compositing materials or the surrounding media [33, 34], which changes the effective refractive index of all the polarization states simultaneously and typically results in the small tuning range of the optical anisotropy. Therefore, the best way to modulate the optical anisotropy of materials, either artificial or natural, is to change the symmetry of their microscopic structures. Research on the liquid crystals has shown good examples of tuning the optical anisotropy via electric or thermally induced microstructures reconfiguration, which can be further applied to metamaterials [35]. In recent works, micromachined tunable metamaterials are reported to have reconfigurable meta-molecules via mechanical actuation [36], which can tune the meta-molecules' geometry without breaking the meta-molecule's symmetry along the actuation direction. The active control of optical anisotropy can be realized by selectively changing the effective refractive index of one polarization state, which results in the tuning of the optical anisotropy between positive and negative values.

References

1. Hecht E (2012) Optics. Pearson Education India
2. Jenkins FA, White HE (1957) Fundamentals of optics. Indian J Phys 25:265–266
3. Lipson A, Lipson SG, Lipson H (2010) Optical physics. Cambridge University Press
4. Mueller JPB, Leosson K, Capasso F (2014) Polarization-selective coupling to long-range surface plasmon polariton waveguides. Nano Lett 14(10):5524–5527. https://doi.org/10.1021/nl501860r

5. Roy JE, Shafai L (1996) Reciprocal circular-polarization-selective surface. IEEE Ant Propag Mag 38(6):18–33. https://doi.org/10.1109/74.556517
6. Sanz-Fernández J, Saenz E, Maagt PD (2015) A circular polarization selective surface for space applications. IEEE Trans Ant Propag 63(6):2460–2470.https://doi.org/10.1109/TAP.2015.2414450
7. Born M, Wolf E (2013) Principles of optics: electromagnetic theory of propagation, interference and diffraction of light. Elsevier
8. Wu S-T, Efron U, Hess LD (1984) Birefringence measurements of liquid crystals. Appl Opt 23(21):3911–3915. https://doi.org/10.1364/AO.23.003911
9. Bartholin E (1669) Erasmi Bartholini Experimenta crystalli Islandici disdiaclastici: Quibus mira & insolita refractio detegitur. Sumptibus Danielis Paulli, Hafniæ
10. Fresnel AJ (2001) Plane waves and wave propagation
11. Fresnel AJ, In Sénarmont HH, In Verdet E, In Fresnel LF (1866) Euvres complètes d'Augustin Fresnel. Imprimerie impériale, Paris
12. Aramaki S, Roy R (1962) Revised phase diagram for the system Al_2O_3—SiO_2. J Am Ceram Soc 45(5):229–242. https://doi.org/10.1111/j.1151-2916.1962.tb11133.x
13. Cardona M, Harbeke G (1965) Optical properties and band structure of Wurtzite-type crystals and rutile. Phys Rev 137(5A):A1467–A1476. https://doi.org/10.1103/PhysRev.137.A1467
14. Ghosh G (1999) Dispersion-equation coefficients for the refractive index and birefringence of calcite and quartz crystals. Optics Commun 163(1):95–102. https://doi.org/10.1016/S0030-4018(99)00091-7
15. Ha S, Tang Y, van Oene MM, Janissen R, Dries RM, Solano B, Dekker NH (2019) Single-crystal rutile TiO_2 nanocylinders are highly effective transducers of optical force and torque. ACS Photonics 6(5):1255–1265. https://doi.org/10.1021/acsphotonics.9b00220
16. Hlubina P, Urbańczyk W (2005) Dispersion of the group birefringence of a calcite crystal measured by white-light spectral interferometry. Meas Sci Technol 16(6):1267–1271. https://doi.org/10.1088/0957-0233/16/6/005
17. Cotter MA (1977) Hard spherocylinders in an anisotropic mean field: a simple model for a nematic liquid crystal. J Chem Phys 66(3):1098–1106. https://doi.org/10.1063/1.434044
18. Yang KH, Rosenblatt C (1983) Determination of the anisotropic potential at the nematic liquid crystal-to-wall interface. Appl Phys Lett 43(1):62–64. https://doi.org/10.1063/1.94121
19. Correas-Serrano D, Alù A, Gomez-Diaz JS (2017) Plasmon canalization and tunneling over anisotropic metasurfaces. Phys Rev B 96(7):075436. https://doi.org/10.1103/PhysRevB.96.075436
20. Della Valle G, Hopkins B, Ganzer L, Stoll T, Rahmani M, Longhi S, Cerullo G (2017) Nonlinear anisotropic dielectric metasurfaces for ultrafast nanophotonics. ACS Photonics 4(9):2129–2136. https://doi.org/10.1021/acsphotonics.7b00544
21. Yu N, Capasso F (2014) Flat optics with designer metasurfaces. Nat Mater 13(2):139–150. https://doi.org/10.1038/nmat3839
22. Zhu WM, Liu AQ, Bourouina T, Tsai DP, Teng JH, Zhang XH, Zheludev NI (2012) Micro-electromechanical Maltese-cross metamaterial with tunable terahertz anisotropy. Nat Commun 3(1):1274. https://doi.org/10.1038/ncomms2285
23. Hsiao H-H, Chu CH, Tsai DP (2017) Fundamentals and applications of metasurfaces. Small Methods 1(4):1600064. https://doi.org/10.1002/smtd.201600064
24. Shi Z, Zhu Alexander Y, Li Z, Huang Y-W, Chen Wei T, Qiu C-W, Capasso F. Continuous angle-tunable birefringence with freeform metasurfaces for arbitrary polarization conversion. Sci Adv 6(23):eaba3367. https://doi.org/10.1126/sciadv.aba3367
25. Tseng ML, Lin Z-H, Kuo HY, Huang T-T, Huang Y-T, Chung TL, Tsai DP (2019) Stress-induced 3D chiral fractal metasurface for enhanced and stabilized broadband near-field optical chirality. Adv Opt Mater 7(15):1900617. https://doi.org/10.1002/adom.201900617
26. Qin J, Deng L, Kang T, Nie L, Feng H, Wang H, Bi L (2020) Switching the optical chirality in magnetoplasmonic metasurfaces using applied magnetic fields. ACS Nano 14(3):2808–2816. https://doi.org/10.1021/acsnano.9b05062

27. Hu J, Zhao X, Lin Y, Zhu A, Zhu X, Guo P, Wang C (2017) All-dielectric metasurface circular dichroism waveplate. Sci Rep 7(1):41893. https://doi.org/10.1038/srep41893
28. Tang Y, Liu Z, Deng J, Li K, Li J, Li G (2020) Nano-Kirigami metasurface with giant nonlinear optical circular dichroism. Laser Photonics Rev 14(7):2000085. https://doi.org/10.1002/lpor.202000085
29. Zhu WM, Liu AQ, Zhang W, Tao JF, Bourouina T, Teng JH, Kwong DL (2011) Polarization dependent state to polarization independent state change in THz metamaterials. Appl Phys Lett 99(22):221102. https://doi.org/10.1063/1.3664131
30. Sheng W (2006) Origins of optical anisotropy in artificial atoms. Appl Phys Lett 89(17):173129. https://doi.org/10.1063/1.2370871
31. Xu J, Li R, Wang S, Han T (2018) Ultra-broadband linear polarization converter based on anisotropic metasurface. Opt Express 26(20):26235–26241. https://doi.org/10.1364/OE.26.026235
32. Chen H-T, Taylor AJ, Yu N (2016) A review of metasurfaces: physics and applications. Rep Prog Phys 79(7):076401. https://doi.org/10.1088/0034-4885/79/7/076401
33. Boardman AD, Grimalsky VV, Kivshar YS, Koshevaya SV, Lapine M, Litchinitser NM, Shalaev VM (2011) Active and tunable metamaterials. Laser Photonics Rev 5(2):287–307. https://doi.org/10.1002/lpor.201000012
34. Turpin JP, Bossard JA, Morgan KL, Werner DH, Werner PL (2014) Reconfigurable and tunable metamaterials: a review of the theory and applications. Int J Ant Propag 2014:429837. https://doi.org/10.1155/2014/429837
35. Zhao Q, Kang L, Du B, Li B, Zhou J, Tang H, Zhang B (2007) Electrically tunable negative permeability metamaterials based on nematic liquid crystals. Appl Phys Lett 90(1):011112. https://doi.org/10.1063/1.2430485
36. Fan K, Padilla WJ (2015) Dynamic electromagnetic metamaterials. Mater Today 18(1):39–50. https://doi.org/10.1016/j.mattod.2014.07.010

Chapter 6
Tunable Chiral Metasurfaces

6.1 Introduction of Chirality

A chiral structure does not superimpose with its mirror image, which is widely found in natural materials [1, 2]. The chiral structure and its imaging structure are named a pair of enantiomers. The chirality exists in many biomolecules with asymmetric molecule structures. The word "chiral" originated from the Greek word for hand because the right and left hand (the 3-D structure) do not superimpose with each other, which is a good example to illustrate the chirality [3].

In 1811, Frech physicist Dominique Arago observed optical activity in chiral materials. He observed that a linearly polarized light beam has a polarization rotation when it passes through a quartz material's optic axis. The ability of a material to rotate the direction of polarization is named optical activity [4, 5]. Other materials, e.g., sugar, tartaric acid, and turpentine, also have chiral molecules which lead to optical activities. In terms of polarization rotation direction, the optical activity can rotate the polarization direction of the incident wave clockwise or counterclockwise, which are noted as dextrorotatory and levorotatory, respectively. The dextrorotatory and levorotatory depends on the structure of the molecules, which does not superimpose with its mirror imaging.

In 1825 Fresnel found that the linear polarization EM wave is the superposition of in-phased right-circularly-polarized and left-circularly-polarized light with the same amplitude. The incident light transmitted through a chiral material remains linearly polarized with a rotation polarization angle θ when the phase delay induced by the absorption is trivial. This polarization angle rotation is known as circular birefringence, which is due to the real part of the refractive index. The difference in the refractive indices, both real and imaginary, for two circularly polarized light comes from the polarization-dependent interaction with the chiral materials. The couplings between the electrical and magnetic fields are opposite in chiral molecules or structures for different circular polarized light. The refractive indices are different for right-circularly-polarized and left-circularly-polarized EM waves due to the interactions between the electromagnetic waves and the chiral molecules or structures.

W. Zhu and A.-Q. Liu, *Metasurfaces: Towards Tunable and Reconfigurable Meta-devices*, Microfluidics and Nanophotonics: Science and Engineering 1,
https://doi.org/10.1007/978-981-19-6925-6_6

Chiral metasurfaces are composed of an array of meta-molecules with a chiral structure, which attracts wide research attention due to many novel optical functions, including realizing a negative refractive index [6, 7]. The negative refractive index material refers to a material with negative permittivity and permeability at the same electromagnetic frequency. Metamaterials with negative permittivities and permeabilities working in microwave regimes are widely demonstrated with chiral meta-molecules of millimeter size. However, negative-refractive-index materials are quite difficult to be realized in the visible regime that covers wavelengths from 400 to 780 nm due to the saturation of the magnetic response at optical frequencies in the meta-molecules small as tens of nanometers. On the other hand, chiral metamaterials and metasurfaces are proposed to obtain the negative refractive indices from a strong electromagnetic coupling between the chiral meta-molecule and the incident waves [8]. A chiral metamaterial can possess the negative refractive index at one circular polarization state when the absolute value of the wavevector is strong enough. A chiral meta-molecule model with a helix coil structure is theoretically proposed for the negative refractive index [9]. Many indefinite chiral metamaterials and metasurfaces are studied, and the backward propagation is investigated with both isotropic and anisotropic meta-molecules [10, 11]. The evanescent wave enhancement and subwavelength focusing are also predicted using a bi-isotropic chiral media [12, 13]. The negative refractive index is both analytically and numerically proposed using different chiral meta-molecules [14, 15]. In addition, the experimental demonstration of the negative refractive index is achieved by using a metamaterial with a 3D chiral structure [16].

Apart from the negative refractive index, the chiral metamaterial can also realize much stronger optical activity than natural materials due to the giant electromagnetic coupling in the meta-molecule. A pioneer work observes a polarization rotation up to 30° in the diffracted light from a planar chiral metamaterial. Later, optical activity is obtained from a non-diffracted planar chiral metamaterial [17, 18]. This optical activity is based on the symmetric breaking of the meta-molecules due to the existence of the substrate. A metasurface with 3D metal screw hole arrays is then proposed for stronger optical activity [19]. Other 3D chiral metasurfaces are also reported with giant optical activities [20–22]. However, optical metasurfaces based on 3D meta-molecules are challenging to be fabricated due to the small feature sizes and complex structures.

On the other hand, two-layered planar structures, which are more easily fabricated, can be designed to have strong optical activities [23, 24]. Achiral metasurfaces can also realize the optical activities with oblique incidence, which is referred to as extrinsic chirality. Giant optical activities are successfully demonstrated by metasurfaces working in the microwave [25], THz [26], infrared, and visible range [27]. In addition, the tunable optical activity is also studied based on chiral metasurfaces [28].

Nonreciprocal transmission in optics is similar to the asymmetric flowing electric current through an electric diode in opposite directions with the same electric bias, which is essential for signal processing. Asymmetric optical transmission of chiral metasurfaces is demonstrated without breaking the reciprocity, which is due to the differences in cross-polarized transmissions between the forward and backward incidence [29, 30]. Double-layered chiral metasurfaces are also studied to enhance asymmetric transmission [31]. The asymmetric transmission of linearly-polarized incidence is also demonstrated by using both multi-layered chiral metasurfaces [32] and metasurfaces working at multiple frequency bands [33].

In conclusion, the chiral metasurface advances the capability to realize negative refractive index, giant optical activity, asymmetric transmission, etc. In this chapter, a chiral metasurface based on semi-3D meta-molecules is discussed, and a tunable GHz chirality is demonstrated with microfluidic metasurfaces, which have high potential applications in the fields of optical imaging, detection, and signal processing, just to name a few.

6.2 Metasurfaces with Semi-3D Structures

Although successfully demonstrated in the relative low-frequency regime, the negative medium is usually hard to be obtained in the optical frequency range through the 3D chiral structure due to the fabrication difficulties. Multi-layered planar chiral metamaterials are proposed and experimentally demonstrated the negative refractive index [34], which still relies on a challenging vertical assembly process for fabrication. Achiral metasurfaces have optical activities with elegantly chosen incident angles, which are highly dependent on incident angles and have limited applications.

On the other hand, the meta-molecules of different structures have been proposed to realize intrinsic chirality, which include the quasi-2D chiral structures, the 3D chiral structures, and the multi-layered chiral structures. The quasi-2D chiral structures rely on chiral structures with the symmetric breaking induced by the substrate, which has weak optical activities.

In this section, A semi-3D meta-molecule is discussed, which depress a planar chiral structure in its normal direction. The meta-molecule is intrinsically chiral and derives strong optical activity from a structured substrate. The metal pattern of the semi-3D meta-molecule designed is a single-layered structure for simplified fabrication processes.

The semi-3D gammadion meta-molecule is a chiral structure that does not superimpose with any of its mirror images, as illustrated in Fig. 6.1. The chirality of the meta-molecules comes from the structured substrate. As a result, meta-molecules based on such structures can be designed to have strong optical activities. In contrast, a planar gammadion meta-molecule is not a chiral structure without considering the symmetric-breaking effect by the substrate, as shown in Fig. 6.1b. The meta-molecules with the semi-3D gammadion structures are compatible with the 2D CMOS fabrication processes. The gammadion meta-molecules are designed to work in the THz frequency region, and the lattice constant of the metasurface is 300 μm. The substrates are patterned with trench structures with a side length s of 170 μm and a depth h up to 50 μm. The metal structures of the meta-molecules are illustrated with yellow color, as shown in Fig. 6.1a, which is an achiral structure without considering the substrate. The trenched substrate folds the metal structures, which, as a result, become the chiral structures with strong optical activities, as shown in Fig. 6.1b.

The transmission spectra of the semi-3D gammadion metasurface are numerically calculated under circularly polarized incidence with the incident frequency ranging from 0.4 THz to 0.6 THz. The right and the left circularly polarized transmissivity, t_{++} and t_{-} respectively, are compared at different h where no differences are observed between t_{++} and t_{-} when $h = 0$ at the investigated frequency range. It is worth pointing out that the planar gammadion on a substrate is also a chiral structure due to the symmetric breaking induced by the substrate. However, the optical activity induced in this design over the investigated frequency range is weak and not observable.

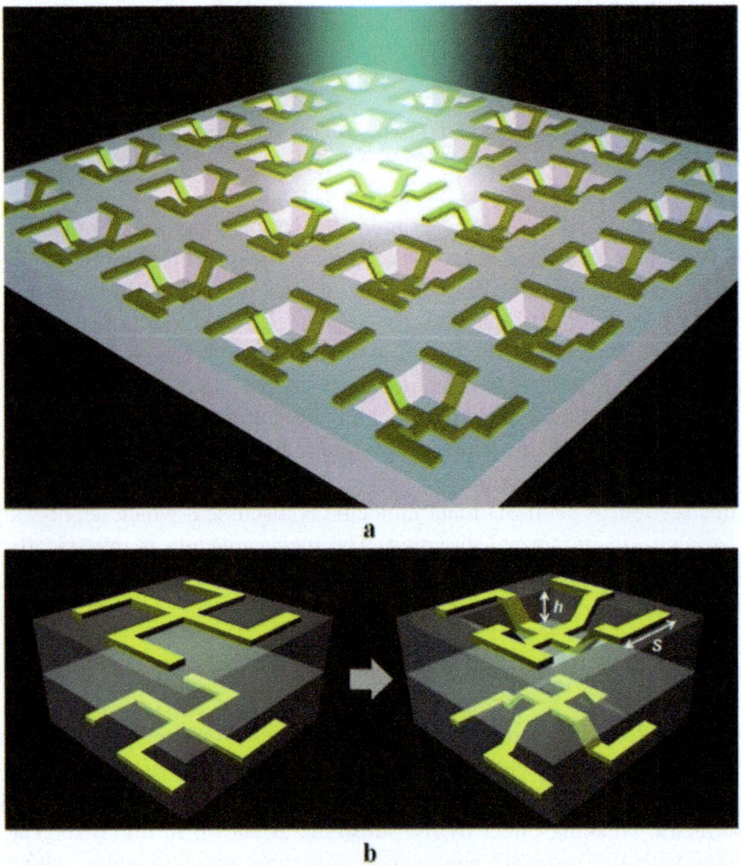

Fig. 6.1 Schematic of the semi-3D chiral metasurface and its design principles **a** an overview of the metasurface **b** the transformation between a planar gammadion meta-molecule and a semi-3D structure

The value of the polarization rotation angle θ and ellipticity tan χ of the semi-3D chiral metasurface is calculated as shown in Fig. 6.2a, b for different trench depths derived from the circular transmittance. The absolute value of the polarization rotation increases as the trench depth h is increasing. The polarization rotation θ increases to 21° at 0.44-THz incident frequency when $h = 10$ μm and increases to approximately 90° when $h = 40$ μm. The same increment occurs when the incident frequency is 0.56 THz. Here, the trench depth is only 1/13 of the incident wavelength, which is deep enough to achieve large polarization rotation angles. Therefore, it proves that a semi-3D gammadion metasurface can realize large polarization rotations with deeply-subwavelength thickness. The ellipticity also becomes non-zero when the depth of the trench h increases, which, however, is not proportional to the trench depth due to its effects on both the amplitude and the phase of the incident THz waves.

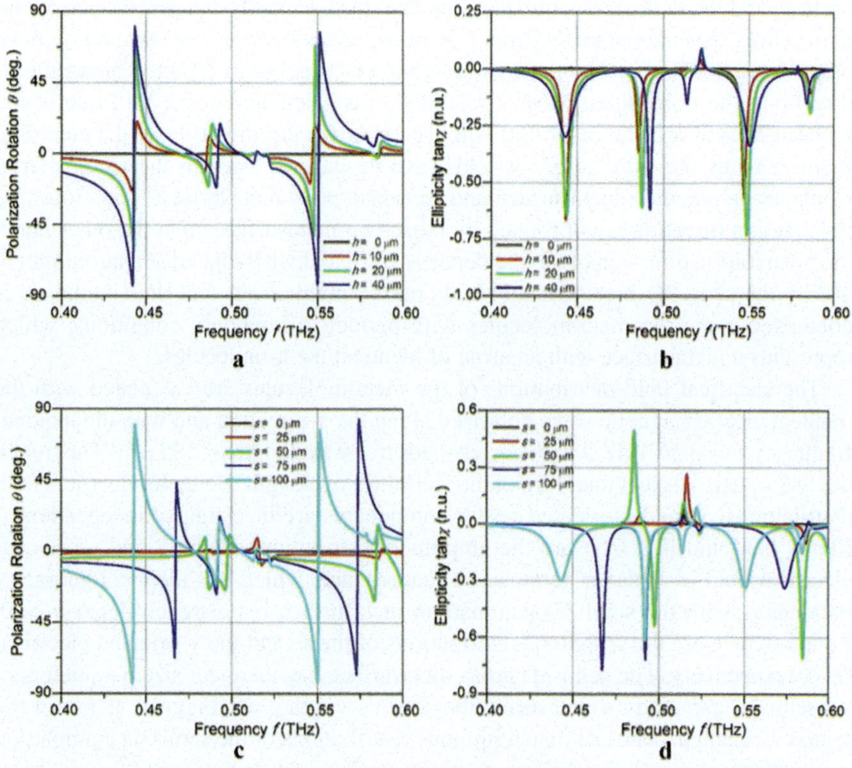

Fig. 6.2 Numerically calculated polarization rotations and ellipticities of the semi-3D gammadion metasurfaces. **a** and **b** show the polarization rotation and ellipticity with different trench depth h, while **c** and **d** show those of different beam lengths s

The optical activity depends on the trench depth h and the geometry of the metal structures, e.g., the beam length s, as shown in Fig. 6.1b. For example, the meta-molecule is an achiral cross structure when $s = 0$. Figure 6.2c, d show the polarization rotation and ellipticity spectra of the semi-3D metasurface with different arm lengths s. Both the polarization rotation and the ellipticity are trivial across the calculated spectrum at $s = 0$, which indicates the metasurface has no optical activity. The polarization rotation and ellipticity can be observed when s is increasing. The polarization rotation angle at 0.52-THz incidence is approximately 10° when the arm length increases to 25 μm. For 0.44-THz incidence, the polarization rotation angle is approximately 90° due to the strong interactions between the electromagnetic waves and the chiral meta-molecules. The designed meta-molecule requires both stereo deflection and planar chirality to achieve the optical activity.

Numerical simulation of the polarization rotation within the azimuth plane is shown in Fig. 6.3 for the planar-gammadion metasurface, the deflected-cross metasurface, and the deflected-gammadion metasurface. Linearly-polarized electromagnetic waves are incident on the front side of the metasurfaces, as shown in Fig. 6.3a, when the 0.56-THz incident waves have the electrical field oriented along the x-direction. The transmitted waves' polarization is calculated at the backside of the metasurfaces at a distance of 450 μm. The polarization direction of the electrical fields remains the same after the EM wave propagates through the metamaterial of planar-gammadion metasurface and trenched-cross metasurface. Therefore, the polarization direction is not rotated by those two metasurfaces. On the other hand, the polarization direction of the incidence is anti-clockwise rotated at approximately 40° by the semi-3D metasurfaces. It should be pointed out that the simulation is conducted using the meta-molecules with periodical boundary conditions, which represents a metasurface with an array of identical meta-molecules.

The electrical field distributions of the meta-molecules are calculated with the incident electromagnetic wave polarized along the x-direction and with the incident frequency $f = 0.56$ THz, which corresponds to a wavelength of 535 μm. The trench depth $h = 40$ μm is less than 1/10 of the working wavelength. Here, the electrical field distributions' x- and y-oriented vector components are investigated independently. Figure 6.3d and Fig. 6.3g are the amplitude distributions of the x and y-oriented electrical field in a planar-gammadion metasurface, which has identical geometry parameters with the semi-3D gammadion metasurface but a trench depth $h = 0$. Figure 6.3e, g are the amplitude distributions of the x- and the y-oriented electrical field, respectively. The semi-3D cross metasurface has identical size parameters to the semi-3D gammadion metasurface but a arm's length $s = 0$. Figure 6.3f, i show the x- and y-oriented electrical field amplitude distributions of the semi-3D gammadion metasurface, respectively. According to the amplitude distributions of the electrical field at the bottom of the substrate, the y-oriented electrical field is much weaker than that of the x-oriented electrical field for both the planar-gammadion metasurface and the semi-3D cross metasurface, which shows that the polarization conversion efficiency of those two structures is trivial. On the other hand, the y and x-oriented electrical fields are comparable for the semi-3D gammadion metasurface, which

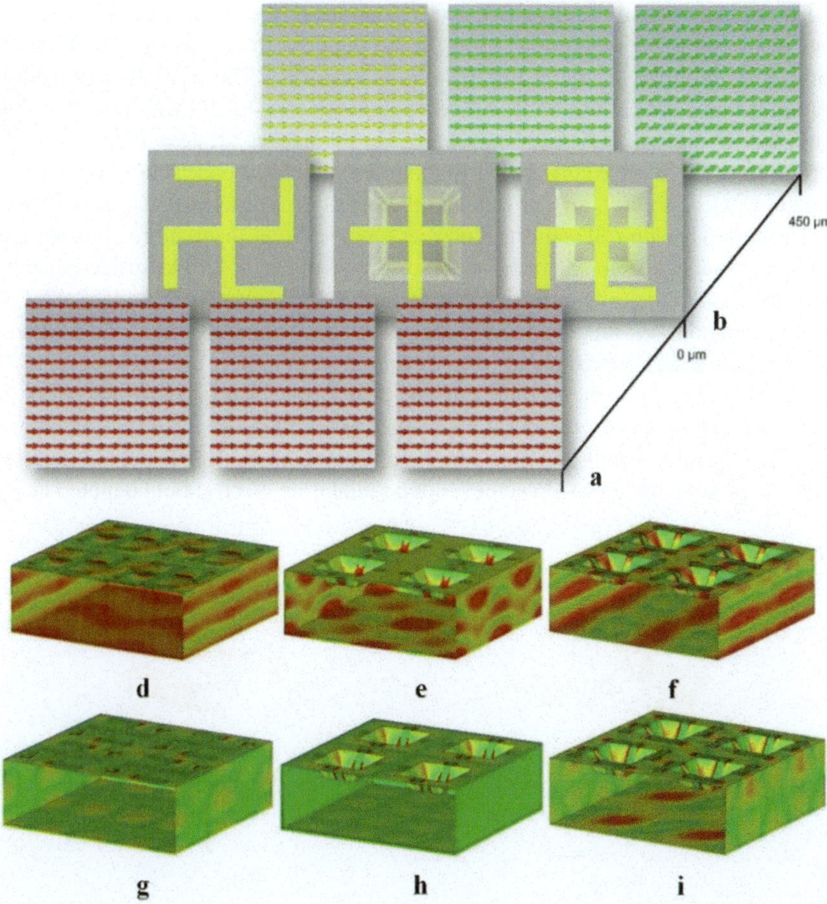

Fig. 6.3 The working principles and magnetic distributions of the semi-3D metasurface. The electrical field of the incident THz wave is oriented along the x-direction with a frequency of 0.56 THz. **a** A linear polarized incident electric fields propagate through **b** a planar-gammadion structure (left column), a semi-3D cross structure (middle column), and a semi-3D gammadion structure (right column), and **c** transmitted electric fields. Electrical field distribution on **d, g** planar gammadion structure and **e, h** semi-3D cross structure and **f, i** semi-3D gammadion structure in **d, e, f** x-orientation and **g, h, i** y-orientation with x-polarization

shows a highly-efficient polarization conversion in the meta-molecule with four-folded rotational symmetry. Therefore, the metasurface with semi-3D gammadion meta-molecules has a strong optical activity due to the chiral structure of the metal wires.

The distributions of the electrical fields show that a semi-3D metal wire structure can be designed with a strong polarization conversion of the incident THz waves. As a result, optical activity can be designed for arbitrary incident frequency by changing the geometry parameters of the meta-molecules.

The optical chirality is investigated by comparing the surface current distributions of the semi-3D meta-molecule with the incident frequency of 0.47 THz and 0.56 THz, which are shown in Fig. 6.4a, b, respectively. The semi-3D gammadion meta-molecules are chosen to have the trench depth $h = 40\ \mu\text{m}$ while the polarization rotation angle of 0.47 THz and 0.56 THz incidences are $\theta = 0°$and $\theta = 75°$, respectively. The surface current distributions are calculated under linearly-polarized incidence, while the red arrows indicate the direction of the instantaneous surface current flow. The semi-3D gammadion meta-molecule has a structure consisting of two orthogonal deflected “Z” metal wires, as shown in Fig. 6.4c. The surface currents on the two arms and the center strip flow along with the same routine under 0.47-THz incidence on both “Z” metal wires. The coupling between the electrical and magnetic resonances is trivial, which results in a weak optical activity.

On the other hand, the surface current distributions on the two arms flow in a routine in the opposite direction along the center strip for 0.56-THz incidence, which can be further interpreted in Fig. 6.4c. The incident electric fields induce an electric dipole on the meta-molecule, which radiates an electric field denoted as E_p.

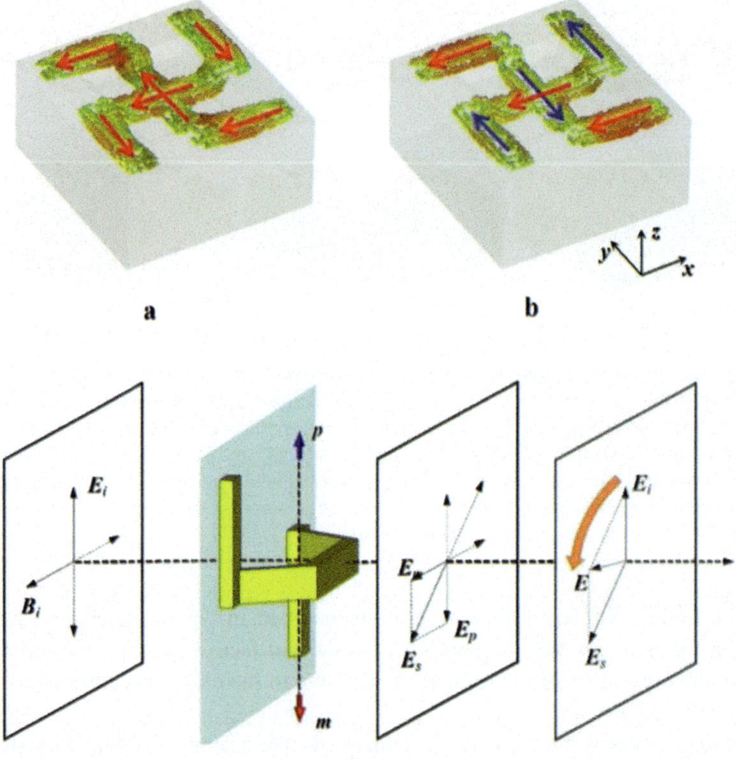

Fig. 6.4 Surface current distributions on the semi-3D gammadion meta-molecule at **a** 0.47 THz and **b** 0.56 THz. **c** The schematic of radiation with the deflected “Z” metal wires

Meanwhile, a magnetic dipole is also induced on the deflected "Z" metal wire due to the spiraling surface current, similar to that in a helix structure but with only one turn. The magnetic dipole is either parallel or antiparallel with the electric dipole. Both oscillating dipoles reradiate electromagnetic waves with orthogonal electric fields denoted as E_p and E_m. The total scattering field E_s combines the two orthogonal reradiated fields, which is not in the same direction as that of the incident electrical field E_i due to the non-zero E_m. As a result, the transmitted light with the "$E_i + E_s$" field is rotated.

The semi-3D gammadion metasurface is fabricated on a single crystal silicon substrate using the wet-etching processes. The squared trench is first etched on the silicon substrate. Then, SiO_2 is deposited and patterned on the substrate, which opens the wet-etching window for the trench etch on the silicon substrate. The wet-etching processes use potassium hydroxide (KOH) solution for etching the silicon substrate at an etching rate of 1 μm/min along the vertical direction. In contrast to the DRIE processes, the wet-etching processes are anisotropic, which etches the silicon substrate both vertically and horizontally. In the fabrication processes, the inversed trapezoid body is etched on the silicon substrate with an angle of 54.7°, which is a constant value defined by the crystal lattice structure of the silicon. The lift-off process is applied to pattern the metal gammadion wires of the meta-molecules. The photoresist (PR) SU8 is uniformly spin-coated on the trenched substrate with a thickness of 5 μm after a 2-min Argon clean process, which enhances the adhesion between metal wires and the silicon substrate. Aluminum (Al) is then deposited on the silicon substrate with a thickness of 0.5 μm after the PR is exposed and developed. The lift-off process is then conducted by using acetone solution with the ultrasonic process to accelerate interactions between the PR and acetone. The semi-3D gammadion meta-molecules are formed by removing the Al on the PR. The single-layered chiral metasurface has a lattice constant $P = 300$ μm and metal wire width $w = 24$ μm. Different arm lengths s are fabricated including $s = 0$ μm, $s = 50$ μm, and $s = 100$ μm. The SEM graph of the THz metasurface is shown in Fig. 6.5a, which consists of a 66 × 66 meta-molecules array and has a footprint of 2 cm^2. The close-up views of the meta-molecules with different trench depths are shown in Fig. 6.5b, c.

The semi-3D gammadion metasurfaces are experimentally characterized using TeraView Spectra 3000 with transmission spectra ranging from 0.4 to 0.6 THz. The terahertz incidence is linearly polarized in the vertical direction, which is defined as the x-polarization state. During the experiment, dry air is supplied in the measuring chamber to eliminate water in the atmosphere, which has strong absorption in the measurement frequency region. The polarization rotation angles θ and ellipticity χ are derived from the measured transmission spectra and are shown in Fig. 6.6. Similar to the simulation results, no polarization rotation is observed when $h = 0$ at 0.56 THz, proving that the optical activity induced by the symmetry breaking from the flat substrate is trivial. As a result, the optical activity is too weak to be observed. The polarization rotation of the 0.56-THz incidence is measured to be 2.5°, 5°, and 14° when $h = 10$ μm, 20 μm, and 50 μm, respectively, as shown in Fig. 6.6a. At the same time, the tangent value of the ellipticity χ also increases from 0 to 0.03 when

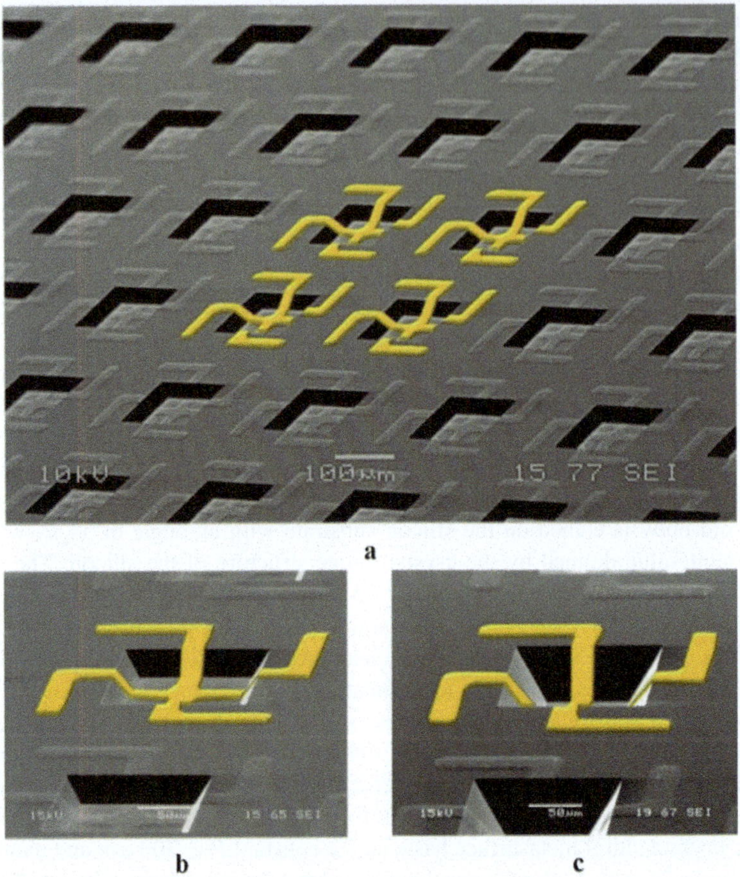

Fig. 6.5 SEM graphs of the semi-3D gammadion metasurface: **a** Overview of the metasurface and close-up views of a single meta-molecule with trench depth of 20 μm (**b**) and 50 μm (**c**). The metal wires are partially highlighted with false yellow color. The metasurface has a footprint of 2 cm^2 with 66 × 66 meta-molecules

h increases from 10 to 50 μm, as shown in Fig. 6.6b. The increment of the optical activities results from the enhanced electromagnetic coupling due to the trenched substrate.

The observed rotation angle is 14° when $h = 50$ μm, less than 1/10 of the working wavelength. The semi-3D gammadion metasurface has demonstrated an optical activity four orders more than that in a quartz crystalline in terms of a rotary power per sample thickness in wavelength unit, which is only 21.7°/mm. The measured optical activity is also 20 times larger than that in the previously reported quasi-2D chiral metamaterial, which has 1° rotation in a sample of 1/6 wavelength thick and is comparable with the optical activities reported in some bilayered structures with 28° rotation in a sample of 1/30 wavelength in thickness.

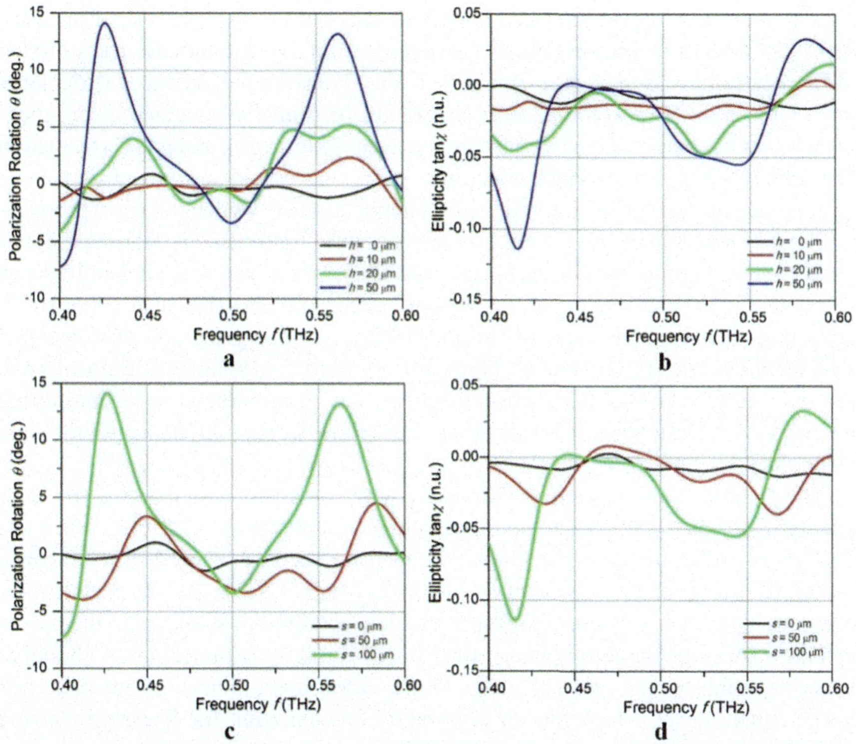

Fig. 6.6 Experimental results of polarization rotation angles (**a**, **c**) and ellipticity (**b**, **d**) of the semi-3D gammadion metasurfaces with different trench depths h (**a**, **b**) and slab lengths s (**c**, **d**)

Besides, the optical activity can be tuned by simply changing the geometry parameters to achieve a desired optical activity at an arbitrary frequency region. The optical activity is also measured using the meta-molecules with different arm lengths s and a fixed trench depth $h = 50$ μm, as shown in Fig. 6.6c, d. No optical activities are observed when $s = 0$, and the meta-molecule is an achiral structure. The meta-molecule becomes a chiral structure, and weak optical activity is induced when $s =$ 50 μm. The polarization rotation θ increases from 5° to 14° for 0.56-THz incidence when s increases from 50 to 100 μm. It proves that the optical activity can also be designed by changing the gammadion arm length.

6.3 Tunable Chiral Metasurfaces Based on Spiral Structures

Chiral materials have asymmetric transmission originating from their different responses to left and right-circularly polarized light. Asymmetric transmissions for circularly polarized light have now been realized using planar chiral metasurfaces [35, 36]. As discussed in the previous section, the asymmetric transmission of linearly-polarized incidence cannot be realized only by metasurfaces with planar meta-molecules. The asymmetric transmission for linearly-polarized electromagnetic waves has been realized by introducing symmetric broken along the propagation direction using the metasurface structures.

It is a desire for optical communication applications to switch the wave propagation between asymmetric transmission to symmetric transmission. However, previously reported chiral metasurfaces lack tunability, which can be switched between chiral and achiral structures. In this section, an array of metallic dual-spiral structures patterned on both sides of the PDMS layer is discussed and demonstrated with reconfigurability between asymmetric transmission to symmetric transmission. The metasurface is switched between achiral and chiral structures by controlling the metal spiral structures from planar to three-dimensional (3D) using microfluidic technology.

The tunable chiral metasurface discussed here consists of a dielectric layer sandwiched between two metallic spiral structures, as shown in Fig. 6.7. When the dual-spiral meta-molecules are planar, the metasurface is an achiral structure since the mirror image can be superimposable on its original. However, the metasurface becomes chiral when the top spiral structure is reconfigured, as shown in the inserts of Fig. 6.7b, since the mirror image of the dual-spiral meta-molecules is no longer superimposable on its original image. Such chiral meta-molecules are patterned into a square array, which has an asymmetric transmission for linearly-polarized electromagnetic waves, as shown in Fig. 6.7a.

The schematic of the tunable chiral metasurface incorporating microfluidic technology is shown in Fig. 6.7b. The period of the metasurface lattice is $P = 5$ mm. The metallic structure (yellow) is designed to be an Archimedean spiral with an inner radius of $R_{\text{in}} = 0.7$ mm, an outer radius of $R_{\text{out}} = 2.0$ mm, a width of $w = 0.3$ mm, and a thickness of 600 nm. The relative twist of the bottom spiral φ_1 is designed to be 0°. The height of the top, bottom PDMS layers and microfluidic channels are $h_1 = 0.5$ mm, $h_2 = 1$ mm, and $h_3 = 80$ μm, respectively. The diameter D of the air reservoir within the PDMS layer is 4 mm. The original state of the metasurface is planar without air injection, as shown in the top insert of Fig. 6.7b. Such planar metasurface is achiral, which has an identical transmission spectrum under forwarding and backward incidence. When the air is pumped into the reservoir, the top PDMS membrane is expanded and forms a PDMS semi-sphere due to air pressure, while the bottom PDMS layer maintains planar since the bottom membrane is much thicker than that on the top. The expanded PDMS membrane stretches out like a balloon, and the metallic structure becomes a 3D spiral. By applying different air pressure, the height of the PDMS semi-sphere can be controlled, resulting in a reconfigurable process to control the height of the spiral structure h. The symmetric breaking of the two spiral structures induces chirality and asymmetric transmission of the tunable chiral metasurface. As a result, the tunable chiral metasurface can be dynamically controlled from achiral to chiral state, enabling the controllable asymmetric transmission of the dual-spiral metasurface.

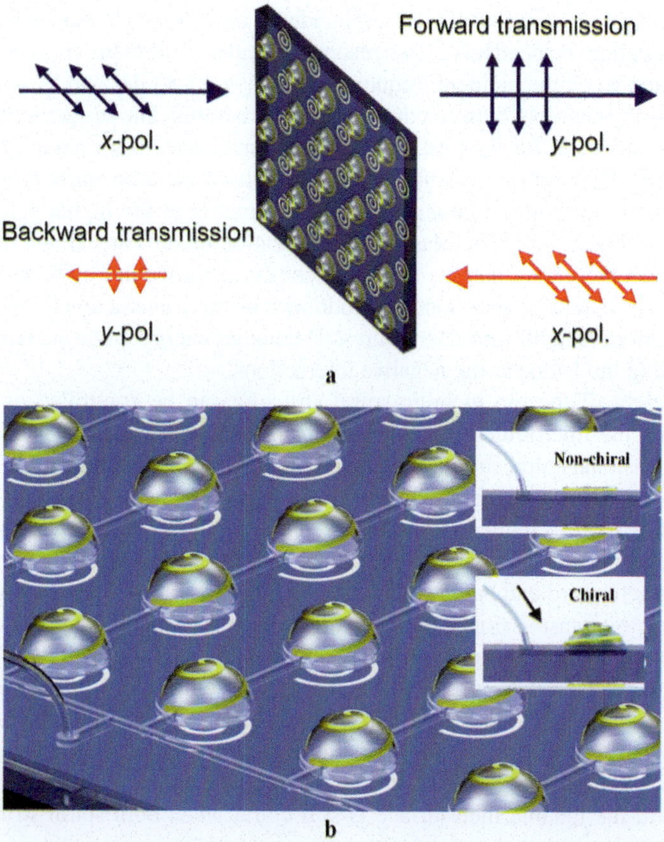

Fig. 6.7 Schematics and working principles of tunable chiral metasurfaces. **a** Illustration of the asymmetric transmission for linearly-polarized waves through tunable chiral metasurface. **b** Dual-spiral structures array incorporating a microfluidic control system. The inserts show the side-views of the dual-spiral meta-molecule's original state and pumped state. The pumping pressure regulates the heights of top spiral structures

The optical responses of the tunable chiral metasurface are calculated by using CST MICROWAVE STUDIO with periodic boundary conditions. The dual-spiral structure is patterned on both sides of the PDMS layer, forming a metal/PDMS/metal sandwich-like structure. The coupling between the top and bottom spiral structures defines the meta-molecules' resonances at specific frequencies, which can be tuned to control the chirality of the metasurfaces in terms of asymmetric transmissions.

Simulation results of the forward and backward transmission of co-polarization T_{xx} and T_{yy} and cross-polarization T_{xy} and T_{yx} are shown in Fig. 6.8a, b, respectively, when both spiral structures are planar. Since the metasurface is achiral, the forward and backward transmission are identical, i.e., $T_{yx}^{f} = T_{xy}^{f} = T_{yx}^{b} = T_{xy}^{b}$ (the superscripts f and b indicate microwave incidence under forward and backward propagation direction, respectively). Two resonance peaks occur in the cross-polarization transmission when the incident frequencies are $f_1 = 25.8$ GHz and $f_2 = 37.1$ GHz, induced by the coupling between the two spiral structures. The surface current distributions are identical for forward and backward incidence, as shown in Fig. 6.8c, d, respectively. The surface currents of top and bottom metallic spiral structures are asymmetric at the lower frequency f_1 and symmetric at the higher frequency f_2, as shown in Fig. 6.8c–f. The plasmon hybridization model can explain the interaction between these two metallic spiral structures, as shown in Fig. 6.8g. The lower energy mode ω_- represents a bonding mode with positive interaction between the top and the bottom metallic spiral structures. The higher energy mode ω_+ represents an anti-bonding mode due to the negative interactions.

The height of the top metallic spiral structure can be controlled by pumping the air into the microfluidic channels with different air pressures, which forms a three-dimensional chiral structure of the meta-molecule. A bonding mode in the two metallic spiral structures is induced at the lower frequency of $f_1 = 25.9$ GHz, while an anti-bonding mode is induced at the higher frequency of $f_2 = 35.8$ GHz. The strength of the surface current distributions indicates that the coupling strength is different between the forward and backward incidence for both resonance peaks. As a result, the asymmetric transmission of the metasurface can be tuned by changing the distance between the two metallic spiral structures. The asymmetric transmission parameter for linearly-polarized waves reaches 5%. The tunable chiral metasurface is controlled from achiral to chiral, which results in a controllable asymmetric transmission.

The dual-spiral wire structures are patterned asymmetrically on both sides of the PDMS substrate to enhance the asymmetric transmission, as shown in Fig. 6.9a, b. As a result, the tunable metasurface is still chiral when both spiral structures are planar.

The heights of the top PDMS layer h_1, bottom PDMS layer h_2, and PDMS channel h_3 are 0.5 mm, 0.42 mm, and 0.08 mm, respectively, where $h_1 - h_3 = h_2$. The PDMS membrane heights of the top and bottom spiral structures are the same. Therefore, both spiral structures are deformed with an equal height of h when air is pumped into the microfluidic channel. The simulation results of the transmission coefficients under forward and backward transmission are used to derive the asymmetric transmission parameters Δ, which are the differences in transmission coefficients between the forward and backward incident waves. When the substrate is planar, i.e., $h = 0$ mm, the metasurface exhibit chirality due to the asymmetric spiral patterns on the top and the bottom of the PDMS substrate. The forward and backward transmission are different, as shown in the black line of Fig. 6.9c. Due to the plasmon hybridization, two resonance peaks are observed at 28 GHz and 37 GHz. The two resonance peaks separate when the spiral's height increases from 0 to 1.4 mm. The coupling between top and bottom spiral wire structures is weakened due to the increase in their spacing. The asymmetric transmission parameter Δ as a function of incident frequency is shown in Fig. 6.9d, which goes up to 50%. The asymmetric transmission exhibits large tunability at the frequency of 37 GHz, which can be tuned from 50% to 0 when the height of the PDMS substrate is changed from 0 to 1.4 mm, as shown in Fig. 6.9d.

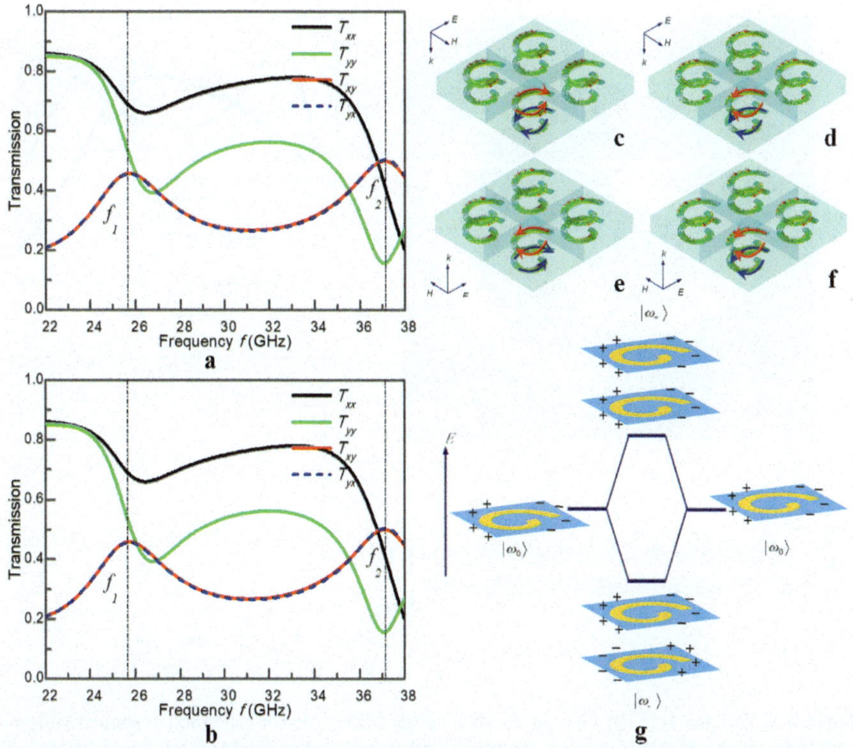

Fig. 6.8 The transmission spectra surface current distributions and quantum analog of the tunable chiral metasurface. The transmission spectra under **a** forward incidence and **b** backward incidence when both layers of spiral structures are planar. **c**, **d** Show the surface current distribution simulation results on the planar meta-molecules at the first peak with frequency f_1 and the second peak f_2, respectively, under forward incidence. Simulation results of the surface current distributions when both layers of spiral structures are planar at **e** the first peak f_1 and **f** second peak f_2 under backward incidence. **g** The energy-level diagram is applied to describe the physics principle of the plasmon hybridization in the tunable chiral metasurface, which is due to the interaction between the top and bottom spiral plasmons. A positive interaction lowers the overall energy of the hybridized mode and results in a bonding mode ($\omega -$), while a negative interaction results in an anti-bonding mode ($\omega +$)

The tunable chiral metasurface is fabricated using photolithography and E-beam evaporation, as shown in Fig. 6.10. The spiral structures are deposited on both sides of the PDMS substrates using E-beam evaporation processes, which are widely explained elsewhere [37]. A shadow mask with the designed spiral patterns is used for metal deposition. Firstly, a 3-nm thickness of chrome film is deposited on the PDMS substrate to enhance the adhesion between the gold and the PDMS. A 650-nm thickness of the gold film is then deposited on the PDMS substrate by using E-beam evaporation. The metal pattern is defined by the shadow mask covered on the PDMS substrate during deposition. The PDMS substrate is flipped over for the deposition on the other side when the shadow mask is well aligned with the previously deposited

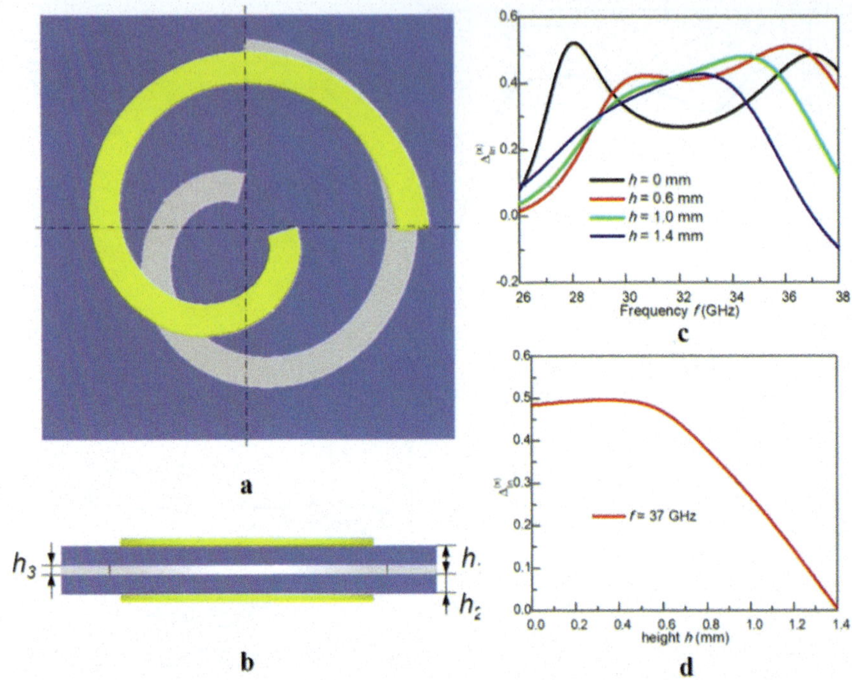

Fig. 6.9 **a** Top and side (**b**) view of the dual-spiral meta-molecule patterned asymmetrically on both sides of the PDMS substrate. Simulation results when the height of the spiral height h is changed from 0 to 1.4 mm. **c** Asymmetric transmission parameter Δ. **d** Asymmetric transmission as a function of spiral height h at the incident frequency of 37 GHz

metal pattern under the microscope. Likewise, a 3-nm chrome film and a 650-nm gold film are deposited in sequence on the PDMS substrate by using E-beam evaporation.

The fabricated metasurface consists of an array of 24 × 24 meta-molecules with a 5-mm period and a total footprint of 120 mm × 120 mm. The original state of the spiral structures is planar, as shown in Fig. 6.10b when the pumping pressure is zero. The top PDMS membrane is expanded, leading to a tunable height of the spiral structure when the air pressure is pumped into the microfluidic channels, as shown in Fig. 6.10c. The tunable chiral metasurface can also be bent into arbitrarily curved surfaces due to the flexibility of the PDMS substrates.

The measurements of the tunable chiral metasurfaces are carried out in an anechoic chamber room for the characterization of RF antennas. The experimental results of tunable chiral metasurfaces are shown in Fig. 6.11 when two spiral structures are asymmetrically patterned on both sides of the PDMS substrate. The asymmetric patterns of the meta-molecules result in the chirality of the tunable metasurface when the PDMS substrates are planar. Therefore, asymmetric transmission under forward and backward incidence is observed, as shown in the black line of Fig. 6.9a, b. The differences between the forward and backward transmission are dramatically

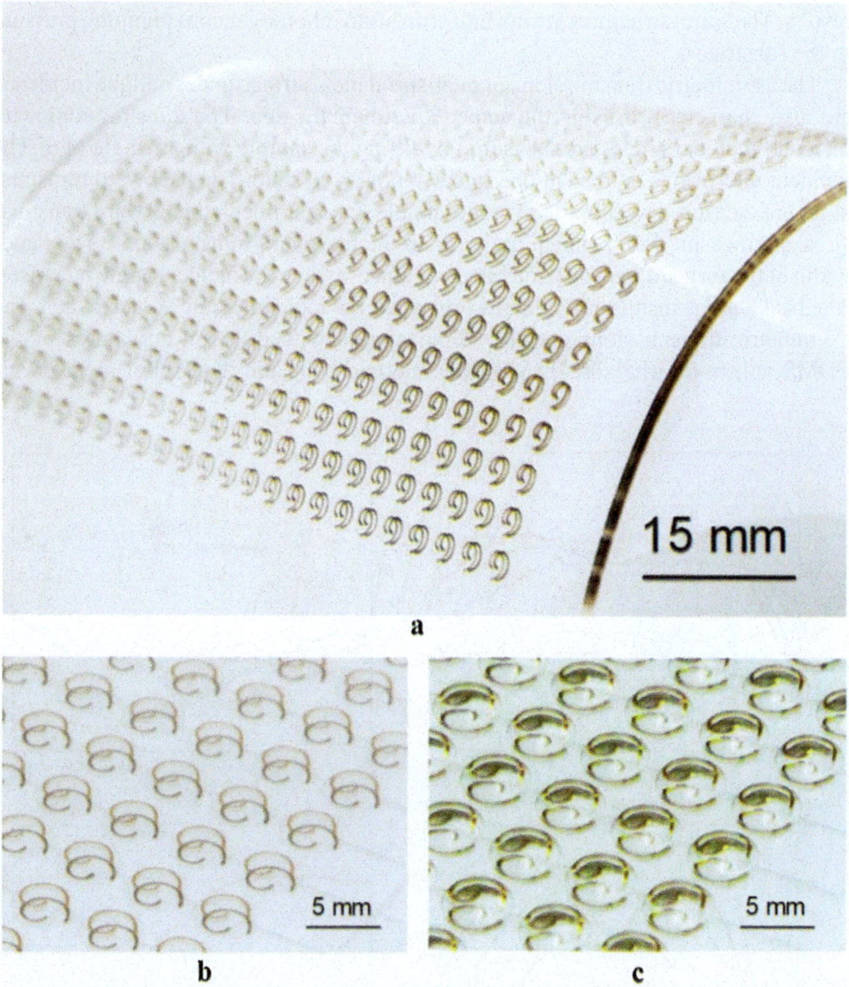

Fig. 6.10 Photographs of the tunable chiral metasurfaces. **a** A full view of the tunable chiral metasurface with two spiral structure layers deposited on both sides of the PDMS. **b** Zoomed-in view of planar dual-spiral metasurface without air pumping when the height of the PDMS structure $h = 0$ mm. **c** The height of the PDMS structure is tuned to h = 1.5 mm by pumping air into the PDMS substrate

enhanced compared with those of the meta-molecules with symmetrically patterned spiral structures. The asymmetric transmission can reach 50%, as shown in Fig. 6.11c. At the frequency of 37 GHz, the metasurface has large tunability on asymmetric transmission, as shown in Fig. 6.11d, where the asymmetric transmission is tuned from 50% to 0 as the height of the PDMS substrates varies from 0 to 1.4 mm. Fabrication errors mainly induce slight mismatches between the simulation and experimental results. The spiral structures are not uniform due to inhomogeneous pumping pressure in the substrates.

The asymmetric transmissions of dual-spiral metasurface under oblique incidence are also characterized using the same experimental setup. The transmissions with different incident angles are measured by tilting the sample with an angle of θ. The incident microwave is fixed at one linearly-polarized state. The backward transmissions are measured by flipping over the sample. In experiment, the forward transmissions are measured to be much higher than the backward transmissions. The bandwidth of the forward transmission becomes narrow when the incident angle increases. The backward transmission remains low when the incident angle is increasing. The asymmetric transmissions are also demonstrated with different curvatures of the PDMS substrate. The soft PDMS substrate can bend the dual-spiral metasurface

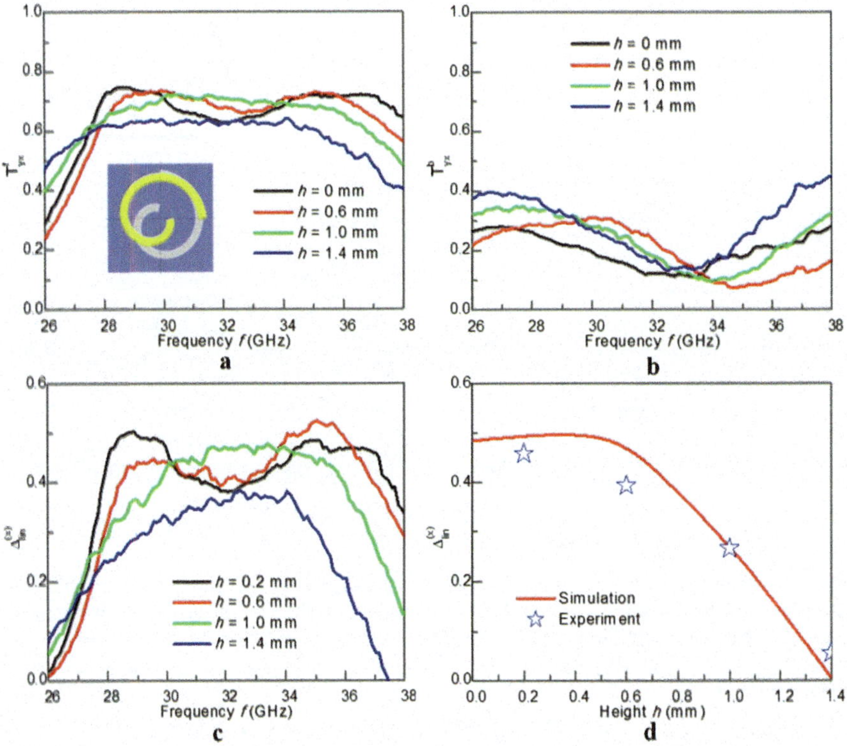

Fig. 6.11 Experimental results of the transmission coefficients with the asymmetrically spiral pattern when the height of the spiral structures varies from 0 to 1.4 mm under **a** the forward incidence and **b** the backward incidence. Experimental results with asymmetrically spiral patterns when the height of the spiral structures h varies from 0 to 1.4 mm. **c** Asymmetric transmission parameter Δ as a function of the incident frequency **d** Asymmetric transmission as a function of h at a fixed incident frequency of 37 GHz

into different curvatures. In experiment, the tunable chiral metasurface is demonstrated with large tolerance of the PDMS substrate curvature, which has potential applications in smart coating and wearable sensors.

6.4 Summary

This chapter focuses on the excitation and manipulation of chiral metasurfaces. Two different metasurfaces are discussed with optical activity induced by different mechanisms. The first metasurface has strong optical activity induced by its semi-3D chiral meta-molecules with structured silicon substrates. The intrinsic chirality is induced by the trenched substrates of single-layered planar chiral meta-molecules. Different trench depths alter the optical activity. The semi-3D metasurface design realizes giant optical activity with simplified fabrication processes.

The second metasurface is designed based on the tunability of the chiral meta-molecules with the substrate tuned from 2 to 3D structures. The meta-molecules consist of dual-spiral structures that interact strongly, which induce two resonance peaks resulting from the plasmon hybridization. The fabrication processes for solid metal incorporating microfluidic technologies supply the chiral metasurface with a large tuning range using polymer PDMS substrate. Experiments prove that the developed dual-spiral metasurface controls the asymmetric microwave transmission by changing the 3D geometries of the meta-molecules. A large tunability of the asymmetric transmission is demonstrated from 50% to 0. In addition, the high asymmetric transmission is observed under oblique incidence from 0° to 45°. The asymmetric transmission maintained at a high level with a broad bandwidth is realized under different curvatures of the curved dual-spiral metasurface.

References

1. Cahn RS, Ingold C, Prelog V (1966) Specification of molecular chirality. Angew Chem, Int Ed Engl 5(4):385–415. https://doi.org/10.1002/anie.196603851
2. Bliokh KY, Nori F (2011) Characterizing optical chirality. Phys Rev A 83(2):021803. https://doi.org/10.1103/PhysRevA.83.021803
3. Harris AB, Kamien RD, Lubensky TC (1999) Molecular chirality and chiral parameters. Rev Mod Phys 71(5):1745–1757. https://doi.org/10.1103/RevModPhys.71.1745
4. Ma X, Pu M, Li X, Guo Y, Gao P, Luo X (2017) Meta-chirality: fundamentals, construction and applications. Nanomaterials 7(5). https://doi.org/10.3390/nano7050116
5. Forbes KA, Andrews DL (2021) Orbital angular momentum of twisted light: chirality and optical activity. J Phys: Photonics 3(2):022007. https://doi.org/10.1088/2515-7647/abdb06
6. Zhou J, Dong J, Wang B, Koschny T, Kafesaki M, Soukoulis CM (2009) Negative refractive index due to chirality. Phys Rev B 79(12):121104. https://doi.org/10.1103/PhysRevB.79.121104
7. Zhao R, Zhang L, Zhou J, Koschny T, Soukoulis CM (2011) Conjugated gammadion chiral metamaterial with uniaxial optical activity and negative refractive index. Phys Rev B 83(3):035105. https://doi.org/10.1103/PhysRevB.83.035105

8. Zhang S, Park Y-S, Li J, Lu X, Zhang W, Zhang X (2009) Negative refractive index in chiral metamaterials. Phys Rev Lett 102(2):023901. https://doi.org/10.1103/PhysRevLett.102.023901
9. Gansel Justyna K, Thiel M, Rill Michael S, Decker M, Bade K, Saile V, Wegener M (2009) Gold helix photonic metamaterial as broadband circular polarizer. Science 325(5947):1513–1515. https://doi.org/10.1126/science.1177031
10. Kwon D-H, Werner DH, Kildishev AV, Shalaev VM (2008) Material parameter retrieval procedure for general bi-isotropic metamaterials and its application to optical chiral negative-index metamaterial design. Opt Express 16(16):11822–11829. https://doi.org/10.1364/OE.16.011822
11. Hooper DC, Mark AG, Kuppe C, Collins JT, Fischer P, Valev VK (2017) Strong rotational anisotropies affect nonlinear chiral metamaterials. Adv Mater 29(13):1605110. https://doi.org/10.1002/adma.201605110
12. Pendry JB (2000) Negative refraction makes a perfect lens. Phys Rev Lett 85(18):3966–3969. https://doi.org/10.1103/PhysRevLett.85.3966
13. Chen L, He S, Shen L (2004) Finite-size effects of a left-handed material slab on the image quality. Phys Rev Lett 92(10):107404. https://doi.org/10.1103/PhysRevLett.92.107404
14. Li Z, Zhao R, Koschny T, Kafesaki M, Alici KB, Colak E, Soukoulis CM (2010) Chiral metamaterials with negative refractive index based on four "U" split ring resonators. Appl Phys Lett 97(8):081901. https://doi.org/10.1063/1.3457448
15. Liu M, Plum E, Li H, Li S, Xu Q, Zhang X, Zhang W (2021) Temperature-controlled optical activity and negative refractive index. Adv Func Mater 31(14):2010249. https://doi.org/10.1002/adfm.202010249
16. Hur K, Francescato Y, Giannini V, Maier SA, Hennig RG, Wiesner U (2011) Three-dimensionally isotropic negative refractive index materials from block copolymer self-assembled chiral gyroid networks. Angew Chem Int Ed 50(50):11985–11989. https://doi.org/10.1002/anie.201104888
17. Kuwata-Gonokami M, Saito N, Ino Y, Kauranen M, Jefimovs K, Vallius T, Svirko Y (2005) Giant optical activity in quasi-two-dimensional planar nanostructures. Phys Rev Lett 95(22):227401. https://doi.org/10.1103/PhysRevLett.95.227401
18. Bai B, Svirko Y, Turunen J, Vallius T (2007) Optical activity in planar chiral metamaterials: Theoretical study. Phys Rev A 76(2):023811. https://doi.org/10.1103/PhysRevA.76.023811
19. Miyamaru F, Hangyo M (2006) Strong optical activity in chiral metamaterials of metal screw hole arrays. Appl Phys Lett 89(21):211105. https://doi.org/10.1063/1.2392787
20. Toyoda K, Miyamoto K, Aoki N, Morita R, Omatsu T (2012) Using optical vortex to control the chirality of twisted metal nanostructures. Nano Lett 12(7):3645–3649. https://doi.org/10.1021/nl301347j
21. Helgert C, Pshenay-Severin E, Falkner M, Menzel C, Rockstuhl C, Kley E-B, Pertsch T (2011) Chiral metamaterial composed of three-dimensional plasmonic nanostructures. Nano Lett 11(10):4400–4404. https://doi.org/10.1021/nl202565e
22. Asadchy VS, Guo C, Faniayeu IA, Fan S (2020) Three-dimensional random dielectric colloid metamaterial with giant isotropic optical activity. Laser Photonics Rev 14(10):2000151. https://doi.org/10.1002/lpor.202000151
23. Liu H, Genov DA, Wu DM, Liu YM, Liu ZW, Sun C, Zhang X (2007) Magnetic plasmon hybridization and optical activity at optical frequencies in metallic nanostructures. Phys Rev B 76(7):073101. https://doi.org/10.1103/PhysRevB.76.073101
24. Li TQ, Liu H, Li T, Wang SM, Wang FM, Wu RX, Zhang X (2008) Magnetic resonance hybridization and optical activity of microwaves in a chiral metamaterial. Appl Phys Lett 92(13):131111. https://doi.org/10.1063/1.2905285
25. Li Z, Alici KB, Colak E, Ozbay E (2011) Complementary chiral metamaterials with giant optical activity and negative refractive index. Appl Phys Lett 98(16):161907. https://doi.org/10.1063/1.3574909
26. Zalkovskij M, Malureanu R, Kremers C, Chigrin DN, Novitsky A, Zhukovsky S, Lavrinenko AV (2013) Optically active babinet planar metamaterial film for terahertz polarization manipulation. Laser Photonics Rev 7(5):810–817. https://doi.org/10.1002/lpor.201300034

27. Cui Y, Kang L, Lan S, Rodrigues S, Cai W (2014) Giant chiral optical response from a twisted-arc metamaterial. Nano Lett 14(2):1021–1025. https://doi.org/10.1021/nl404572u
28. Ren M, Plum E, Xu J, Zheludev NI (2012) Giant nonlinear optical activity in a plasmonic metamaterial. Nat Commun 3(1):833. https://doi.org/10.1038/ncomms1805
29. Fedotov VA, Mladyonov PL, Prosvirnin SL, Rogacheva AV, Chen Y, Zheludev NI (2006) Asymmetric propagation of electromagnetic waves through a planar chiral structure. Phys Rev Lett 97(16):167401. https://doi.org/10.1103/PhysRevLett.97.167401
30. Singh R, Plum E, Menzel C, Rockstuhl C, Azad AK, Cheville RA, Zheludev NI (2009) Terahertz metamaterial with asymmetric transmission. Phys Rev B 80(15):153104. https://doi.org/10.1103/PhysRevB.80.153104
31. Wu L, Yang Z, Cheng Y, Zhao M, Gong R, Zheng Y, Yuan X (2013) Giant asymmetric transmission of circular polarization in layer-by-layer chiral metamaterials. Appl Phys Lett 103(2):021903. https://doi.org/10.1063/1.4813487
32. Bokhari SHA, Cheema HM (2021) Ultra-wideband, wide angle, asymmetric transmission based chiral metasurface for C and X band applications. Sci Rep 11(1):11724. https://doi.org/10.1038/s41598-021-91126-1
33. Huang CX, Zhang JJ, Wu L, Zhang C, Yang J, Yang LX, Cui TJ (2020) Multi-band tunable chiral metamaterial for asymmetric transmission and absorption of linearly polarized electromagnetic waves. Adv Theory Simulations 3(12):2000179. https://doi.org/10.1002/adts.202000179
34. He M, Han J, Tian Z, Gu J, Xing Q (2011) Negative refractive index in chiral spiral metamaterials at terahertz frequencies. Optik 122(18):1676–1679. https://doi.org/10.1016/j.ijleo.2010.10.024
35. Li Z, Liu W, Cheng H, Chen S, Tian J (2016) Tunable dual-band asymmetric transmission for circularly polarized waves with graphene planar chiral metasurfaces. Opt Lett 41(13):3142–3145. https://doi.org/10.1364/OL.41.003142
36. Zhang F, Pu M, Li X, Gao P, Ma X, Luo J, Luo X (2017) All-dielectric metasurfaces for simultaneous giant circular asymmetric transmission and wavefront shaping based on asymmetric photonic spin–orbit interactions. Adv Func Mater 27(47):1704295. https://doi.org/10.1002/adfm.201704295
37. Chen Y (2015) Nanofabrication by electron beam lithography and its applications: a review. Microelectron Eng 135:57–72. https://doi.org/10.1016/j.mee.2015.02.042

Chapter 7
Tunable Absorber Based on Meta-fluidic-Materials

7.1 Introduction

The meta-fluidic-materials (MFM) refer to the metasurfaces and metamaterials partially made of liquid materials, either the surrounding medium or the meta-molecules. The MFM are tunable metasurfaces or metamaterials, which have reconfigurable electromagnetic properties or functionalities controlled dynamically through external excitations. Different from previously proposed metasurfaces based on rigid substrates, such as silicon, PCB board, glass, etc., MFM wholly or partially consist of soft or elastic materials, such as PDMS [1], polyimide [2], silk [3], and papers [4], which are flexible and offer new control freedoms in terms of mechanical actuation. Reconfigurable metasurfaces based on flexible substrates have been widely demonstrated with working frequencies ranging from visible to RF [5–7]. For example, flexible metasurfaces based on PDMS substrates have tunable electromagnetic properties induced by mechanical actuation, which changes the geometries of the meta-molecules and their lattice constants [8]. Additionally, bendable metasurfaces are realized with extraordinary electromagnetic and mechanical properties, such as curved surfaces [9], negative index materials [10], perfect absorbers [11], filters [12], etc.

Liquid metals at room temperature, such as mercury (Hg) and gallium-indium alloy (eutectic gallium-indium (EGaIn) or gallium-indium-tin (Galinstan)), have fair electric conductivities and low reactivities. The liquid–metal metasurfaces usually use microfluidic technology with microchannels to control and reconfigure the liquid metal. The first liquid–metal-based meta-molecule was demonstrated in 2009 using mercury as the resonant structure [13]. The microfluidic split-ring resonators (MF-SRR) consist of micro PDMS channels and liquid metal (Hg). The input and output channels of the liquid metal are composed of two PDMS fibers. The proposed meta-molecule shows large tunability when the MF-SRR array is switched between the non-filling state and the full-filling state of liquid metal. However, mercury is a toxic material, limiting its applications in developing MFM.

W. Zhu and A.-Q. Liu, *Metasurfaces: Towards Tunable and Reconfigurable Meta-devices*, Microfluidics and Nanophotonics: Science and Engineering 1,
https://doi.org/10.1007/978-981-19-6925-6_7

On the other hand, gallium-indium alloy is a good candidate due to its non-toxic properties, which are also in the liquid state at room temperature. A microfluidic technology demonstrates an EGaIn-based metasurface working in the THz region [14]. It should be pointed out that the gallium-indium alloy can be easily oxidized when exposed to the air. A typical solution to this problem is to dissolve the oxide layer using hydrochloric (HCl) vapor and purge the microchannels with inert gases or other non-reactive gases, such as Nitrogen, Helium, etc. The metasurfaces based on MFM are realized by refilling the liquid metal into the microchannels to form the meta-molecules. As a result, the transmission spectra of the metasurfaces can be significantly tuned by changing the filling factors of EGaIn within the meta-molecules.

The liquid metals at room temperature have many limitations considering the practical applications. For example, mercury is toxic. Gallium-indium alloy is easily oxidized when exposed to the air, wetting the glass, silicon, or PDMS surface. Furthermore, sodium–potassium alloys are highly reactive and flammable. Liquid dielectric materials are then proposed based on water, sodium chloride solution, etc. Those liquid materials are abundant, low-cost, and bio-compatible and have diverse functionalities and promising applications. A theoretical analysis of the water-based metasurface is also proposed, suggesting different tuning mechanisms of the MFM based on water-filled reservoirs, which can be realized by using thermal tuning, mechanical tuning, and gravitational tuning of the magnetic and electric resonances [15]. For example, temperature variations cause the shift of the resonances and the water absorption. As a result, the resonator Q-factor is tuned by thermal effects, which can be realized by various external or internal excitations, such as chemical reactions, microwave radioactivity, etc. Mechanical actuation allows the liquid materials to be reshaped into different geometries, which can be realized by active pumping, introducing air bubbles in water, using ultrasound to vibrate meta-molecules, etc. Gravitation allows liquid material to partially fill containers on a rotating plate and changes their geometries. Water is also applied as a phase shifter between a photonic crystal slab and a dielectric metasurface based on Mie and Bragg resonances [16]. The metasurface is realized by rationally designed microstructures where the resonance wavelength of the Mie mode is higher than that of the Bragg mode.

Conversely, The meta-fluidic-material switches to photonic crystals when the Bragg mode has a longer resonance wavelength than that of the Mie mode. Recently, water droplet arrays have been applied to perfect absorbers working in microwave frequency regions [17]. However, the in-depth physics understanding of water droplets is still limited. The lack of real-time reconfigurability and unreliability induced by gravity limit their practical applications due to the water droplets' arrangements on the substrates in a free-standing style. The water-based metasurfaces are usually demonstrated in the microwave frequency regions since the refractive index is relatively high in the microwave region, making water a perfect candidate for Huygens metasurfaces. The sodium chloride solutions are also applied to MFM in the terahertz region [18]. Two layers of S-shaped microchannels separated by a PDMS layer are designed for the injection of sodium chloride solution. The symmetry is broken by pumping the top and bottom microchannels with sodium chloride solution of different concentrations. As a result, a tunable polarization rotation is realized by using MFM.

7.2 Perfect Absorption Based on Water Resonators

Design of Metasurfaces Based on Water Resonators

The electromagnetic properties of a metasurface are defined by a complex electric permittivity $\varepsilon(\omega) = \varepsilon' + \varepsilon''$ and a magnetic permeability $\mu(\omega) = \mu' + \mu''$. The $\varepsilon(\omega)$ and $\mu(\omega)$ of a metasurface can be rationally designed to achieve the impedance-matching condition to suppress the reflection from the interface between the metasurface and its surrounding media. As a result, a near-unity absorption can be realized by eliminating the transmission by using a reflectionless metasurface with a ground plate. The metal ground plate has similar electromagnetic properties as the perfect electrical conductors (PEC), inducing negligible Ohmic loss. The absorption of the metasurface is mainly due to the dielectric layers of the MFM. As a result, the lossy dielectric materials are good candidates for metasurface absorbers confining the incident electromagnetic energy within the dielectric layers. In this section, an array of water droplets embedded in PDMS micro reservoirs is applied to a metasurface absorber with broadband and near-unity absorption.

An array of meta-molecules based on water-droplets resonators is designed to control the effective permittivity and permeability of the meta-fluidic-material, illustrated in Fig. 7.1a. Each meta-molecule consists of a water-sphere cap sandwiched by a top PDMS layer with $h_2 = 0.1$ mm and a bottom PDMS spacing layer with the height of h_3, which is bonded to a metallic layer. A circular reservoir connected with a microchannel is in the top PDMS layer with a height of $h_1 = 0.08$ mm. The diameter of the circular reservoir is denoted as $D = 4.4$ mm, while the width of the microchannel is denoted as $w = 0.3$ mm, as shown in Fig. 7.1b, c.

The reservoir expends along the vertical direction and forms a water-sphere-cap resonator when pumped with water since the diameter of the reservoir is designed to be much larger than its height. The incident electromagnetic waves can induce both the electrical and magnetic resonances within the water resonator with properly designed D and h, as shown in Fig. 7.1d. Additionally, the impedance matching condition can be achieved to eliminate the reflection of the incidence while the metal ground plate stops the transmission of the incidence. As a result, the meta-fluidic-material functions as a perfect absorber with incident electromagnetic waves confined within and absorbed by the dielectric layer, i.e., the water resonator and the PDMS layer.

The height of the water resonator is defined by the expansion of the PDMS membrane surrounding the micro reservoir, which depends on balancing the pumping pressure of water and the pressure induced by the stretching of the PDMS. A higher water-pumping pressure further expands the PDMS membrane and a water resonator

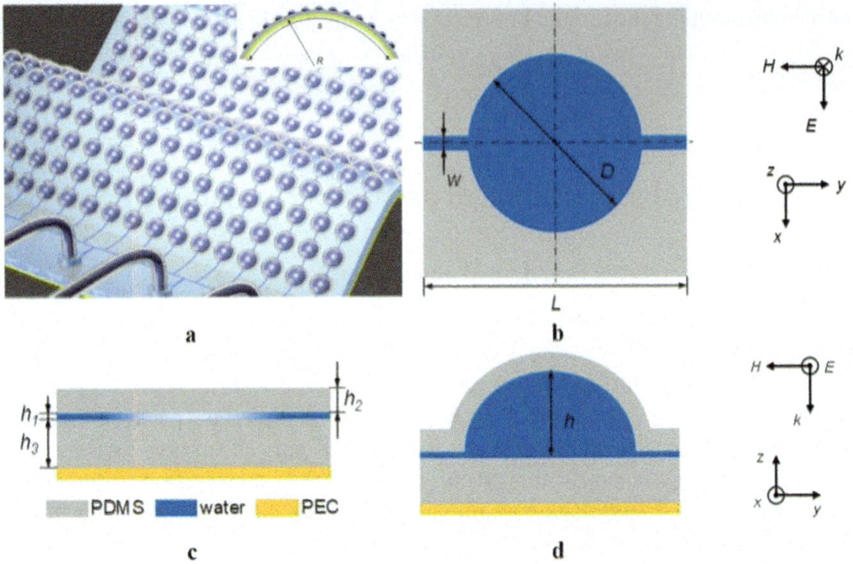

Fig. 7.1 Schematics of the metasurface based on water resonators **a** Overview of the water-resonator-based metasurface with a top PDMS layer and a bottom metal layer. The insert shows a curved metasurface with a radius of *R*. **b** Top view, **c** side view of the empty micro reservoir, and **d** side view of the water resonator with the height of the water resonator *h* that the pumping pressure can control

with a larger height *h*. As a result, the geometry of the water resonator can be controlled by the pumping pressure of the water.

The Numerical Analysis of Water-Resonator Metasurface Absorber

The water-resonator meta-molecule consists of a PDMS/water/PDMS/metal microstructure with four layers. Magnetic resonances can be excited within the meta-molecules at the microwave frequency region due to the high permittivity of the water, which is essential for the impedance matching conditions. In this section, the absorption spectra of the MFM are calculated using the finite element method with different water resonators.

The normalized absorption *A* of the water-resonator meta-fluidic-material is derived from the *S*-parameter S_{11} and S_{21} that $A = 1 - |S_{11}|^2 - |S_{21}|^2$, where S_{11} is the amplitude of the reflected field and S_{21} is the amplitude of the transmitted field, respectively. S_{21} is always zero since the metallic ground plate blocks all incidence transmission.

The absorption spectra of the meta-fluidic-material with different heights of the PDMS layer are shown in Fig. 7.2a. The absorption of the meta-fluidic-material is relatively insensitive to the height of the top PDMS layer, with a height much smaller than the incident wavelength. On the other hand, the absorption spectra are highly dependent on the height of the PDMS h_3 spacing layer between the water resonators and the ground plate. The absorption peaks first arise to near unity at $h_3 = 1.0$ mm and then fall to below 80% at $h_3 = 2.0$ mm when the bottom PDMS layer's height increases, as shown in Fig. 7.2b. The PDMS spacing layer can vary the phase differences between the reflected electromagnetic waves from the metal plate and the water-resonator layer making their interferences either constructive or destructive. However, the interactions between the water resonators and the ground plate are weakened when the PDMS spacing layer is thick. In this design, the height of the top and bottom PDMS layers are chosen to be $h_2 = 0.6$ mm and $h_3 = 1.0$ mm, respectively.

The impedance $Z(f)$ of the meta-fluidic-material can be retrieved from the reflection (S_{11}) and transmission (S_{21}) spectra of the S-parameters considering the meta-fluidic-material an effective homogeneous slab. Here, the electromagnetic waves are set to be normally incident on the surface of the meta-fluidic-material. The normalized optical impedance Z can be written as the function of S-parameters as [19],

$$Z = \pm\sqrt{\frac{(1 + S_{11})^2 - S_{21}^2}{(1 - S_{11})^2 - S_{21}^2}} \tag{7.1}$$

where the real part of the impedance determines the sign in the equation. The calculated impedance spectra $Z(f)$ with $h = 1.6$ mm is shown in Fig. 7.2c. The real part of the impedance Z′ is approximately unity (Z′ ~ 1), and the imaginary impedance Z″ is nearly zero (Z″ ~ 0) at $f_1 = 13.1$ GHz and $f_2 = 36.9$ GHz, showing that the reflection is approximately zero. Simulated results of the meta-fluidic-material absorption spectrum are shown in Fig. 7.2d when the height of the water resonator is fixed at $h = 1.6$ mm. Two peaks are observed within the absorption spectrum with the peak frequencies at $f_1 = 13.1$ GHz and $f_2 = 36.9$ GHz with the absorption of 98% and 99%, respectively. It proves that the near-unity absorption can be achieved at multiple frequencies with perfect impedance match conditions.

Spoof surface plasmon polaritons (SPPs) are observed at the interface between the water-resonator layer and metal ground, as shown in Fig. 7.2e, f. The incident electromagnetic wave is strongly confined in the meta-fluidic-material due to the excitation of the spoof SPPs. As a result, the incidence can be effectively absorbed by the meta-fluidic-material with a high water absorption coefficient in the microwave frequency region. More importantly, the absorption frequency band can be tuned by changing the resonance frequencies of the two peaks, i.e., f_1 and f_2.

Fabrication Results of the Water-Resonator Metasurface

The fabrication of the MFM is based on soft photolithography, which transfers the microstructures from a silicon substrate to a PDMS membrane by using the molding method. First, a 675-μm thick silicon wafer is baked in a 120 °C oven for 15 min after the cleaning process using acetone and IPA. The photoresist (AZ 4620) is spin-coated on the wafer, which is then prebaked to drive off excess photoresist solvent. After prebaking, the silicon wafer with photoresist is covered by a plastic mask with designed microstructures and exposed to UV light, which causes a chemical change of

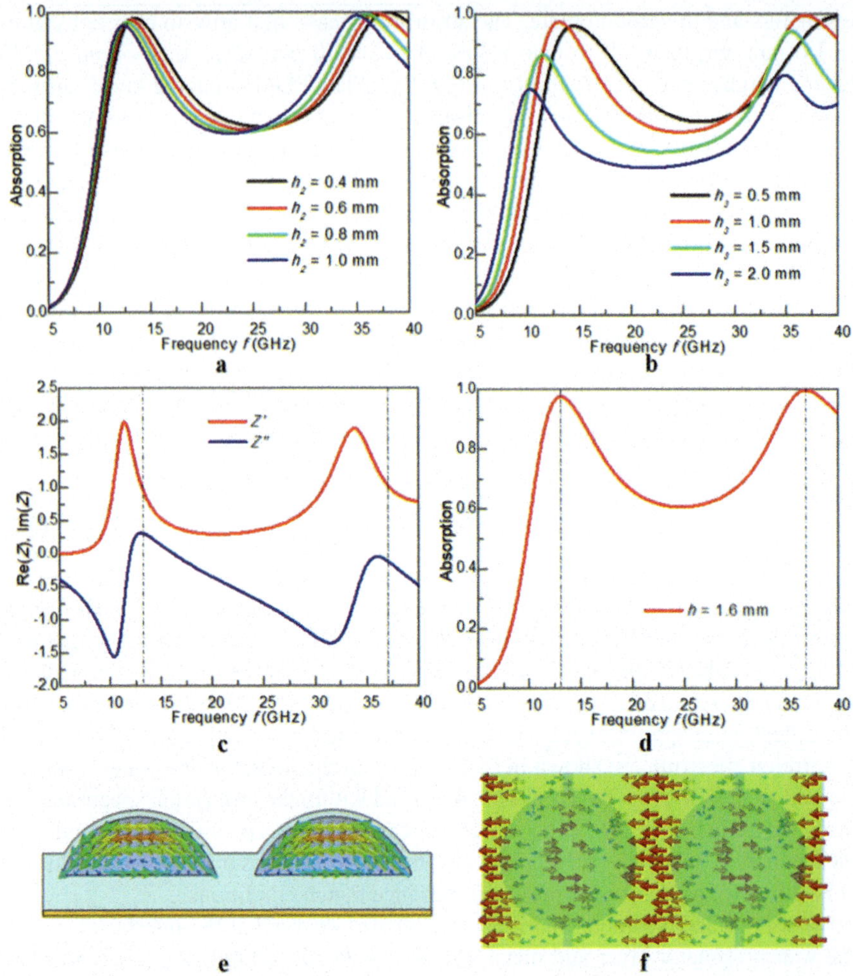

Fig. 7.2 Numerical analysis of the meta-fluidic-material **a** absorption spectra when the height of the top PDMS h_2 varies from 0.4 to 1.0 mm and h_3 is fixed to be 1.0 mm. **b** Absorption spectra when the height of the bottom PDMS h_3 is changed from 0.5 to 2.0 mm and h_2 is fixed to be 0.6 mm. **c** The calculated optical impedance and **d** absorption spectrum when the height of water-resonator h = 1.6 mm. **e** and **f** Show the surface current distribution of the water resonators from the side view and top view, respectively

the photoresist and makes it possible to be removed by the developer. The remaining photoresist protects the underneath silicon from being etched by the plasma chemical agent, which is applied onto the silicon wafer to remove its uppermost layer. The silicone mold is then obtained after removing the remaining photoresist. As a result, the microstructures on the silicon wafer are formed by the photolithography process, which is the complementary structure to that of the PDMS membrane. The alternative fabrication process for the mold is to form the microstructures using the photoresists (SU8), which is discussed in Chap. 3.

The next step is to fabricate the PDMS channel using the silicon mold. First, the PDMS base and curing agent are mixed with the weight proportion of 11:1, which is put in a vacuumizer to expel the air within the PDMS mixture. Then, the PDMS is poured onto the silicon mold and baked in a 75 °C oven for 2 h. The PDMS is then cured, and the designed patterns are transferred to the PDMS membrane by peeling it off it from the silicon mold. The PDMS spacing layer is fabricated by pouring the liquid PDMS mixture onto the flat substrate with a mold regulating its height. A spin-coat process is necessary when a sub-millimeter height is required. The microfluidic channels are formed by bonding the patterned PDMS membrane to the PDMS spacing layer after plasma treatment. This process can be applied to a curved substrate to fabricate the meta-fluidic-material with a curved substrate.

The meta-fluidic-material consists of 24 × 24 meta-molecules based on water resonators with a 5.5-mm period, with a footprint of 132 mm × 132 mm. The pumping pressure of the water can control the height of the water resonators. Figure 7.3a shows the graph of the water resonators with a low pumping pressure, and the height of the water resonators is 0.4 mm. The height of the water resonators is tuned to 1.6 mm by increasing the pumping pressure, as shown in Fig. 7.3b. The water resonators are uniformly controlled by the pumping pressure evenly distributed on each PDMS reservoir due to the mechanisms similar to the communicating vessels with still water inside, which can be derived using Bernoulli's equation. The meta-fluidic-material can be bent to a curved surface with a soft PDMS substrate, as shown in Fig. 7.3c. The microfluidic control system regulates the water within the PDMS reservoirs, which is much more stable than the droplet arrays on a flat substrate with a free-standing style.

The experimental characterization of the meta-fluidic-material is carried out in an anechoic chamber room, as shown in Fig. 7.3d. Two horn antennas are used to transmit and receive the incident and reflected microwaves and are connected to the vector network analyzer (VNA, Agilent N4693A). The antennas are placed on a circular track with the meta-fluidic-material in its center, which has a radius of 1.2 m for far-field measurement. Here, two different antennas are used during the experiment to cover the broad working band of the meta-fluidic-material ranging from 5 to 40 GHz. One is for 5–18 GHz, and the other is for 18–40 GHz. The source antenna can move along the arc left to the meta-fluidic-material with an incident angle of φ_i. In the meantime, the receiver can move along the arc to the right with a reflective angle of $\varphi_\text{r} = \varphi_\text{i}$. For curved metasurface measurement, the source antenna is fixed at the incident angle of 0°, pointing along the symmetry axis of the meta-fluidic-material. The incident wave is similar to a plane wave with the meta-fluidic material placed at its waist. In the experiment, both the reflection spectra with different incident angles are measured.

The Experimental Characterizations of the Water-Resonator Metasurface

The absorption spectra of the meta-fluidic-material are measured with different heights of the water resonators, as shown in Fig. 7.4a. The absorption spectra have

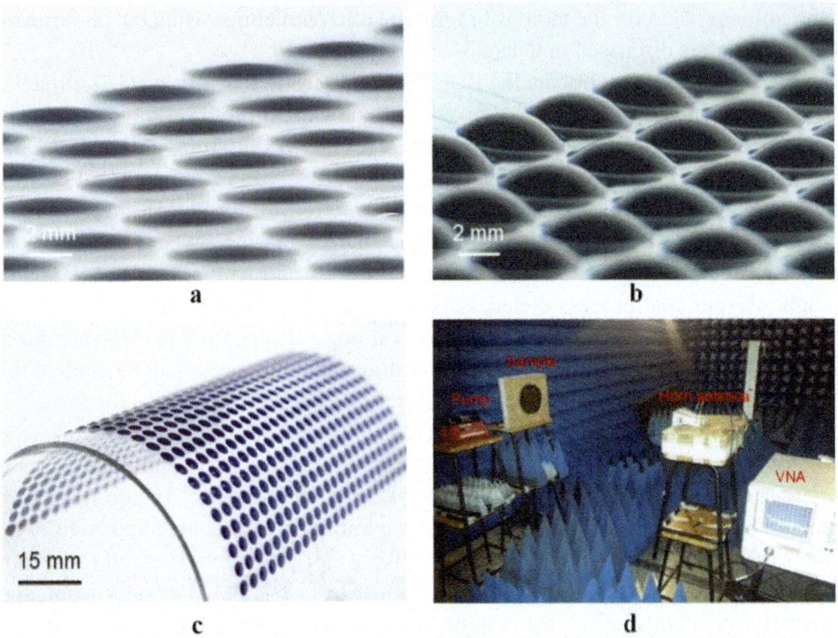

Fig. 7.3 Photographs and experimental setup for MFM based on water resonators. **a** Water is pumped into the PDMS reservoirs with low pumping pressure to form water resonators with a height of $h = 0.4$ mm. **b** The height of the water resonator is increased to $h = 1.6$ mm by increasing the pumping pressure. **c** The metasurface is bent to form a meta-fluidic-material with a curved substrate. **d** Microwave measurement setup in an anechoic chamber room. The horn antennas are connected to the VNA and placed 1.2 m away from the sample to measure the far-field reflection spectra. Two antennas are used during the measurement: one for 5–18 GHz and the other for 18–40 GHz

two peaks, denoted as f_1 and f_2, as shown in Fig. 7.2d. The absorption peak at the lower frequency f_1 shows a redshift from 20 GHz (K band) to 15 GHz (Ku band) when h is increased from 0.4 to 1.2 mm. The magnetic resonances induce this absorption peak within the water resonator. Therefore, f_1 shifts to a lower frequency region as the height of the water resonator increases. The absorption peak at the higher frequency f_2 remains unchanged since it is caused by the grating effect from the periodic resonator array. As a result, the absorption band can be tuned by tuning the magnetic resonance frequency when the water resonator's height is changed. The tunable absorption of the meta-fluidic-material is experimentally demonstrated, as shown in Fig. 7.4a. The maximum absorption efficiency approaches the unity (99%). Here, the absorption band is defined by the frequency region when the absorption efficiency is above 90%, which is measured to be 78.5% of the central frequency with $h = 0.4$ mm. The simulation and experimental results mismatch is less than 2%, indicating a good agreement. In the experiment, the non-uniform water resonators are observed with high pumping pressure, mainly due to the inhomogeneous PDMS membrane. The thickness and mechanical properties of the PDMS membrane can be affected by the fabrication errors, e.g., the inhomogeneous PDMS mixtures, micro air bubbles, the temperature distribution of the baking oven, etc. In addition, the dispersion of the water resonator also affects the absorption bandwidth.

The reflection spectra of all the reflection angles are measured to estimate the scattering properties of the meta-fluidic-material. Here, the source antenna is pointing along the symmetric axis of the meta-fluidic-material with a fixed incident angle of 0°. In the meantime, the receiving antenna is moved along the arc to measure the reflection spectra at different angles. The reflection spectra of an aluminum metal plate are measured as the reference, which shows a significant scattering effect when the reflection angle is smaller than 15°. On the other hand, the scattering is significantly suppressed at all reflection angles when the metallic plate is covered by the meta-fluidic-material, as shown in Fig. 7.4b.

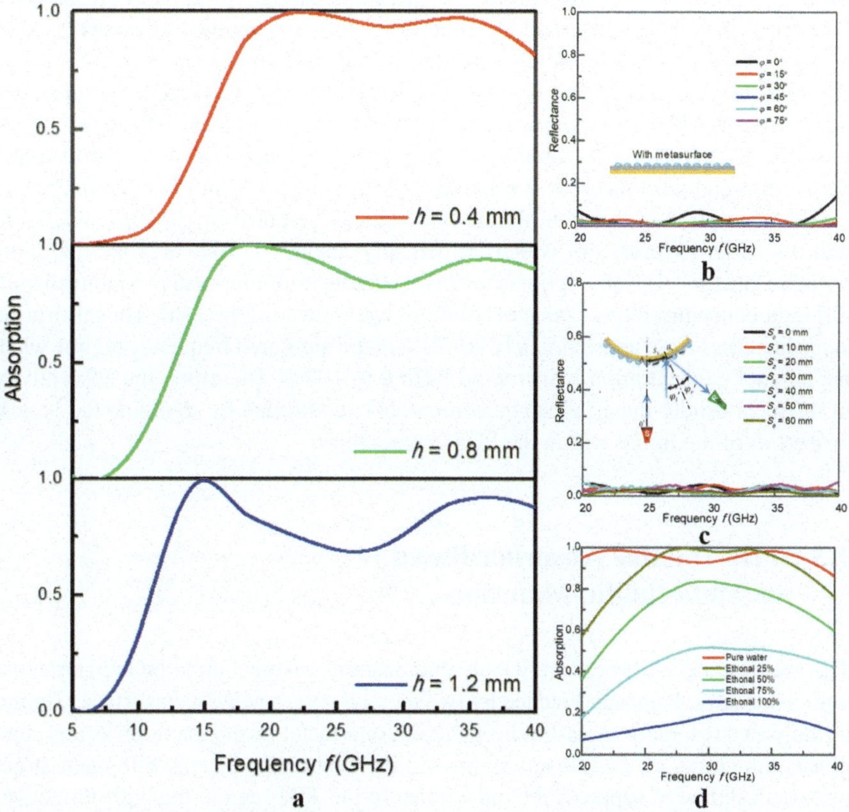

Fig. 7.4 The experimental characterization of the MFM **a** The absorption spectra when the height of the water resonator h varies from 0.4 to 1.2 mm. **b** The reflection spectra of meta-fluidic-material at different reflection angles. The incident angle is fixed at 0°. **c** The reflection spectra of the meta-fluidic-material on a curved substrate with a radius $R = 200$ mm. **d** The broadband absorption tuned by the ethanol solution concentration variation from 0% (pure water) to 100% (pure ethanol). The height of the water resonator is fixed at 0.4 mm

The broadband absorption of MFM with large angular tolerance enables the omnidirectional broadband absorptions on arbitrarily curved surfaces, which is a promising candidate for the stealth coating application. The reflected electromagnetic waves are scattered in different directions when a curved surface is illuminated. Here the scattering properties of the meta-fluidic-material are characterized by measuring its scattering spectra at far-field. The receiving antenna travels throughout the semi-omnidirectional space with the shift denoted as S, as shown in Fig. 7.4c. The incident antenna is fixed at the same position for flat meta-fluidic-material measurement as shown in Fig. 7.4b, which is due to the rotational symmetry of the curved meta-fluidic-material. The incident electromagnetic wave is similar to the plane wave with the meta fluidic material placed at the waist of the incident beam. The measured reflection spectra are normalized by those of the curved copper substrate with the same radium but without the MFM. Figure 7.4c shows the reflection spectra of the MFM with a small curvature of $R = 200$ mm and the height of the water resonator $h = 0.4$ mm. The reflections to all the Omni directions are significantly suppressed ($<$ 10%) from 20 to 40 GHz, which indicates the absorption is larger than 90% due to the meta-fluidic-material. Similar results can also be observed with curved substrates when its radius $R = 100$ mm, which shows the meta-fluidic-material can be applied to substrates with complex geometries. It can be concluded that an omnidirectional and ultra-broadband absorber is achieved by using the meta-fluidic-material.

The absorption of the meta-fluidic-material can be tuned by changing the liquid within the PDMS reservoirs, which dramatically changes the optical properties of the liquid resonators. The absorption spectra of a flat meta-fluidic-material pumped with different concentrations of ethanol solutions are shown in Fig. 7.4d. The absorption level decreases from approximately 90–20% in the measured frequency regime when the ethanol concentration is increased from 0 to 100%. Therefore, the absorptivity of the meta-fluidic-material can be continuously controlled by changing the optical properties of the liquid within the PDMS reservoirs.

7.3 THz Tunable Absorber Based on Meta-fluidic-Materials

The metasurface absorbers are widely demonstrated using a three-layered structure with a dielectric layer sandwiched by two metal layers, a MIM metasurface. Similar to the water-resonator meta-fluidic-material, one metal layer functions as a ground plane eliminating all the transmissions, while the other metal layer with microstructures is designed to suppress the reflections. In the THz region, the high-refractive-index dielectric materials are rare, which are essential for impedance matching based on magnetic resonances. Therefore, meta-molecules with metal microstructures are often proposed for metasurface absorbers where the magnetic resonances can be induced by asymmetric surface current displacements of the top and bottom metal layers. The magnetic responses induced by the MIM metasurfaces are typically highly dependent on the incident frequency, which results in a narrow absorption band. The semi-3D meta-molecules are proposed to have significant magnetic responses, e.g., the vertical split-ring structures, requiring complex fabrication processes. Additionally, the absorber also requires broadband and angular tolerance for practical applications, which are challenging to realize by the 2D meta-molecules.

On the other hand, 3D meta-molecules can be formed and controlled by using microfluidic technologies with multi-layered microchannels. The optical properties of the 3D meta-molecules, i.e., electrical and magnetic resonances, are highly dependent on their geometries and can be controlled by the external pumping systems. The MFM incorporated with microfluidic technology provides great opportunities to reconfigure the 3D meta-molecules, leading to multiple degrees of control freedom. This section discusses a meta-fluidic-material based on liquid metal, with 3D meta-molecules regulated by a microfluidic control system.

The Design of Meta-fluidic-Materials

The working principle of the meta-fluidic-material is shown in Fig. 7.5a, which has four different layers. Two PDMS layers sandwich a silicon layer with penetrated cavities. The liquid metal layer is embedded within the bottom PDMS layer, connecting with silicon cavities via microfluidic channels. The heights of the liquid metal cylinders depend on the air pressure within the air inlet embedded within the top PDMS layer. As a result, the geometry tuning of the meta-molecules becomes three-dimensional, which is different from most reconfigurable metasurfaces. The 3D geometry of the meta-molecules introduces strong magnetic resonances via the coupling between adjacent liquid–metal pillars, which are controlled by the pumping pressure of the air channels.

The liquid metal meta-molecule has a Lorentz-like resonance described by effective parameters, i.e., permittivity $\varepsilon(\omega) = \varepsilon' + i\varepsilon''$ and permeability $\mu(\omega) = \mu' + i\mu''$. Here, ε' and ε" represent the real and imaginary parts of the effective permittivity, respectively. And μ' and μ'' represent the real and imaginary parts of the effective permeability, respectively. The gaps between the adjacent liquid–metal pillars function as the capacitors driven by the external electrical fields. The liquid–metal pillars connect with the bottom liquid–metal reservoir and function as inductive loops when driven by the external magnetic fields. As a result, the meta-molecule, consisting of four liquid–metal pillars, can be described by an equivalent circuit, as shown in Fig. 7.5b. The effective capacitance, inductance, and resistance are denoted as C, L, and R, respectively. The resonant frequency ω of a meta-molecule based on the U-shaped liquid–metal resonator can be described as,

$$\omega = \frac{1}{2\pi\sqrt{\left(L + \mu_r S^2/V\right)C}} \tag{7.2}$$

where μ_r is the permeability of the surrounding medium, $S = Gh$ is the area of the U-shape resonator, and the $V = P^2h$ is the volume of the U-shaped resonator.

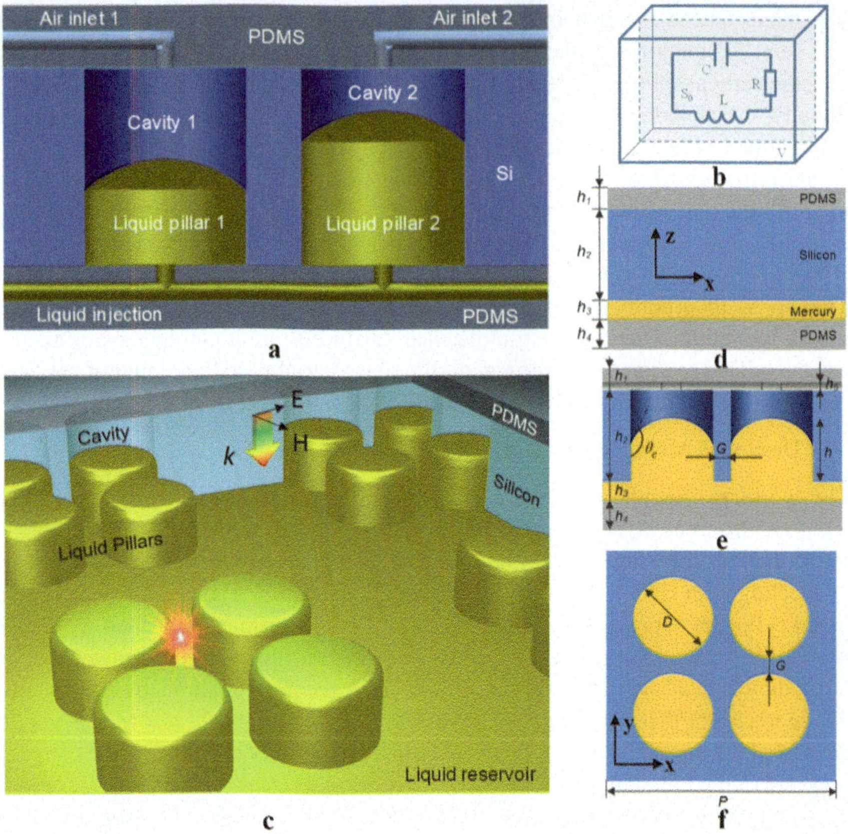

Fig. 7.5 Schematics of meta-fluidic-material based on liquid metal **a** Design of the liquid–metal-based metasurface consists of liquid metal confined in the silicon cavities that can be tuned in the vertical direction by using microfluidic technology. The liquid metal (mercury) is injected from the bottom PDMS layer and fills the silicon cavity. The heights of the liquid–metal pillars can be controlled by applying different air pumping pressures from the microchannels on the top of the silicon cavities. **b** An equivalent circuit of the U-shaped resonator is composed of two connected metal pillars. **c** The schematic of the tunable absorber based on the liquid–metal-material with an array of liquid–metal pillars. **d** The Layout of multi-layered meta-fluidic-material consists of four different layers, including two PDMS layers at the bottom and on the top, respectively, a silicon layer embedded with liquid–metal pillars and a ground-plane layer of liquid metal. The height of the mercury layer h_3 and the bottom PDMS layer h_4 are 0.05 mm and 1 mm, respectively. The schematics for one meta-molecule composed of four liquid–metal pillars are shown in **e** (side view) and **f** (top view). The silicon and top PDMS layer heights are denoted as h_2 and h_1, respectively

Here, G is the gap between two adjacent liquid–metal pillars, h is the height of the liquid–metal pillars, and P is the period of the meta-molecules, as shown in Fig. 7.5d–f. Therefore, the resonance frequency of the meta-molecules can be written as

$$\omega = \frac{1}{2\pi\sqrt{(L + \mu_r Gh/P^2)C}} \tag{7.3}$$

Both the capacitance C and the inductance L are the functions of h, which are controlled by the pumping pressure applied to the air microchannels. As a result, the resonance frequency of the meta-molecule ω can be dynamically controlled by the microfluidic system. For example, the effective capacitance C increases when the liquid–metal pillars h height increases. In the meantime, the effective inductance L increases when h increases, resulting from a larger cross-section area of the inductive loop. According to Eq. 7.3, the resonance frequency of the meta-molecules redshifts when h is increasing and vice versa.

A meta-fluidic-material based on meta-molecules with four liquid–metal pillars is proposed for frequency-agile and wide-angle absorption in the THz region, as shown in Fig. 7.5c. The pillars connect via the liquid–metal layer, forming U-shaped resonators with designable resonances induced by electrical and magnetic fields. The geometry design of the meta-molecules can achieve the optical impedance matching, which is essential for high absorbtivity of the meta-fluidic-material. Furthermore, the meta-molecules are designed to have four identical liquid–metal pillars with four-fold rotational symmetry, making the meta-molecules polarization-independent and supplying the meta-fluidic-material with large angular tolerance for wide-angle absorption. The air inlet is connected to the top PDMS layer, creating a uniform air pressure for the simultaneous and identical tuning of the liquid–metal pillars.

Numerical Analysis of the Tunable Absorber

The polarization states of the incident THz waves are defined by the incident plane of tilted incidence, where transverse electrical (TE) and transverse magnetic (TM) states represent the incidences with electrical and magnetic fields perpendicular to the incident plane, respectively. The same definition of the polarization states is applied to normal incidence for the consistency of the simulation results. Figure 7.6a shows the absorption spectra of TE (red-solid line) and TM (blue dots) polarized normal incidence, which overlapped with each other. The height h of the liquid–metal pillars is 70 μm. The meta-fluidic-material has a polarization-independent absorption due to the four-fold rotational symmetry of the meta-molecules. The insert shows a contour map of the TM-polarized incidence absorption spectra when h ranges from 25 to 100 μm, where the color represents the absorptivities. The absorption peak is at 0.32 THz when $h = 70$ μm, and the near-unity absorption (99%) is achieved, as shown in Fig. 7.6a. The magnetic field distributions induced by TE and TM-polarized incidences are shown in Fig. 7.6b, c, respectively. The areas with overall positive and negative charge distributions are represented by the "+" and "–" symbols, respectively.

The alternating distributions of positive and negative charges form a loop current within the adjacent liquid–metal pillars, which induces a strong magnetic dipole resonance and changes the effective permeability of the meta-fluidic-material. In the meantime, the substantial field enhancement between the gaps of the liquid pillars

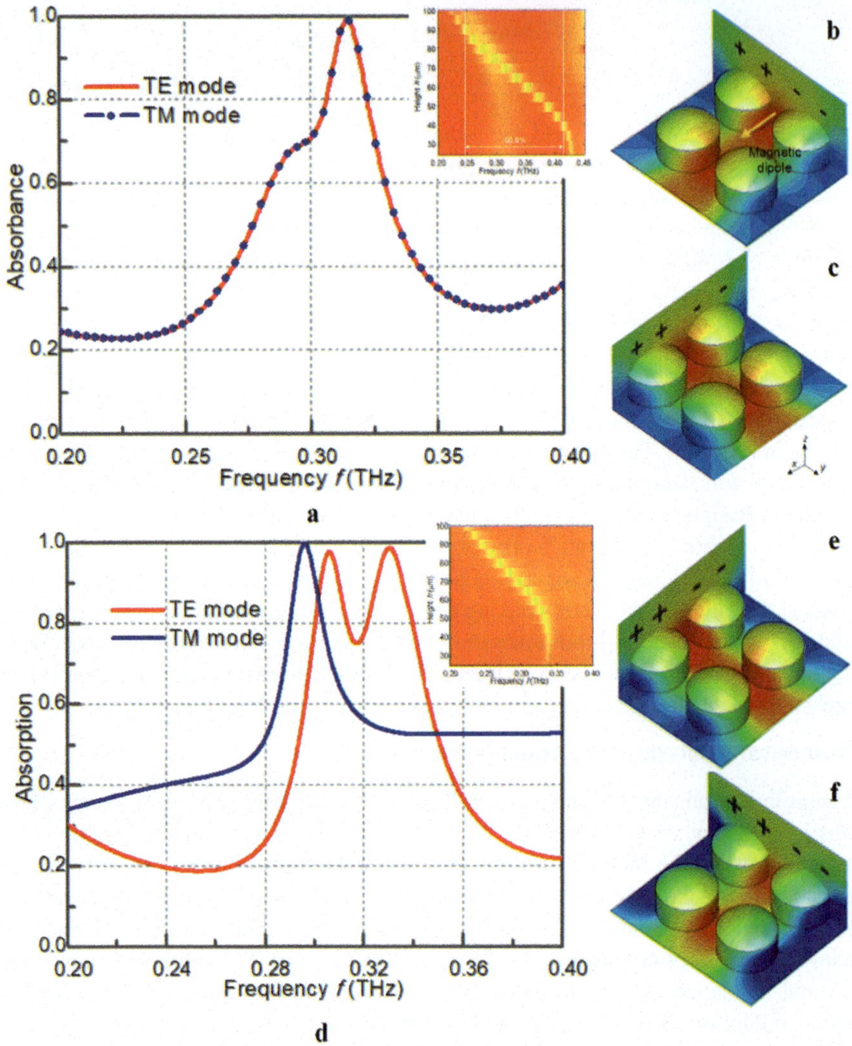

Fig. 7.6 Simulation results of the meta-fluidic-material with normal ((a), (b), and (c)) and tilted ((d), (e), and (f)) incidences. **a** The absorption spectra with TE incidence (red-solid line) and TM incidence (blue dots) when the heights of the liquid–metal pillars are $h = 70\ \mu m$. The insert shows the absorption color map of TM incidence when the heights of the liquid–metal-pillars a tuned from $h = 25\ \mu m$ to $h = 100\ \mu m$. **b** Magnetic field distribution at the peak frequency (0.32 THz) with TE mode. **c** Magnetic field distribution at the peak frequency (0.32 THz) with TM incidence. The incident angle θ is fixed at 60° for tilted incidence **d** Absorption spectra with TE and TM mode when the heights of the pillars keep at 72 μm. The absorption spectra are represented by the red and blue lines for TE and TM-polarized incidences, respectively. The insert shows an absorption color map of tilted incidence ($\theta = 60°$) obtained using the same simulation parameters as that of (a). **e** and **f** The magnetic field distribution with TM and TE incidence at the first resonance peak, respectively

shows the meta-molecules function as micro-capacitors, which changes the effective permittivity of the meta-fluidic-material. As a result, the effective optical impedance of the meta-fluidic-material can be designed by changing the geometry of the meta-molecules.

The absorption spectra contour map of the meta-fluidic-material is shown in Fig. 7.6a when the heights of the liquid–metal pillars range from 25 to 100 μm. According to Eq. 7.3, the increment of h results in the redshift of the resonance frequency, which is similar to the absorption peak frequency. The peak frequency of the absorption spectrum is 0.414 THz when the height of the liquid–metal pillar is 40 μm. The absorption peak frequency shifts to 0.246 THz when h increases to 90 μm. The tuning range of the absorption peak reaches 50.9% of the central frequency when h is tuned from 40 to 90 μm. The peak absorptivity is above 90% during the tuning process. Figure 7.6b, c show a rotational symmetry, which explains the polarization-independent absorption of the meta-fluidic-material.

In addition, the tilted incidences are also applied to the meta-fluidic-material to investigate the angular dependence of its absorption spectra. Figure 7.6d–f show the simulation results of the meta-fluidic-material with oblique incidences, which have a fixed incident angle $\theta = 60°$. Figure 7.6d shows that the absorption spectra are changed under tilted illuminance for both TE (red line) and TM (blue line) polarization states, which are no longer overlapping. The TM-polarized incidence has one absorption peak within the simulation frequency region, ranging from 0.2 to 0.4 THz. The tilted incidence results in a redshift of the absorption peak. The magnetic field distribution of the meta-molecule with TM incidence is shown in Fig. 7.6e, which has the identical magnetic resonance mode as the meta-molecules under normal incidence, as shown in Fig. 7.6b, c. The magnetic resonance remains unchanged with different incident angles for TM-polarized incidence since the incident magnetic fields are perpendicular to the incident plane. However, the orientation direction of the incident electrical field varies with the incident angle, which results in a redshift of the electrical resonance. As a result, the optical impedance matching condition is satisfied by a tilted incidence with a lower frequency, and the absorption peak is redshifted.

On the other hand, the absorption peak splits into two when the meta-fluidic-material is illuminated by a tilted incidence with the TE polarization state. Figure 7.6f shows the magnetic field distributions of the meta-molecule excited by TE-polarized incidence with $\theta = 60°$, which indicates a high order magnetic resonance. The adjacent U-shaped resonators coupled with each other induce another magnetic resonance at the side of the meta-molecules since the U-shaped resonators are periodically arranged in the meta-molecule array. The high-order magnetic resonance results in two resonance peaks of the effective permeability, and the optical impedance matching condition can be achieved by the incident THz waves with two different frequencies. As a result, the absorption peak of TE-polarized incidence split into two when $\theta = 60°$.

The absorption contour map of tilted incidence is shown in Fig. 7.6d. Here, the incident polarization state is chosen to be TM for better comparison with the normal incidence. The tuning range of the absorption peak frequency covers 27.6% of the central frequency, which is tuned from 0.25 to 0.33 THz, maintaining the peak absorptivity above 90%. The simulation results show that the meta-fluidic-material is a good candidate for broadband THz absorbers with large tunability.

Fabrication Results and Experimental Characterizations

The fabrication processes of the meta-fluidic-material consist of two major steps, i.e., the DRIE etching for the silicon layer and the soft lithography for the PDMS layer. First, the photoresist AZ 4620 is spin-coated on the silicon, followed by a prebaking process to expel the excess solvent within the photoresist. The prebaked wafer is exposed under UV light with a patterned plastic mask. After developing processes to remove the exposed photoresist, the uncovered part of the silicon wafer is etched with a depth of 120 μm. Then the photoresist is removed by acetone, and the silicon wafer is thinned to 100 μm using the backside grinding process. As a result, the silicon layer is patterned with periodical holes for liquid metal. The PDMS layers are fabricated using soft photolithography, similar to that of the water-based resonators. The master for PDMS channels is fabricated using the SU-8 photoresist spin-coated on the silicon substrate. The stiffness of the PDMS layer for the microchannels is designed to be slightly higher than that of the PDMS reservoirs discussed in the previous section since the PDMS reservoirs have to be deformed during the tuning processes. Here the stiffness of the PDMS layer is controlled by the weight proportion of the base and curing agent for liquid PDMS solutions. The PDMS layers are bonded to the silicon layer with the surface treatment, which enhances the adhesion between PDMS and silicon.

The meta-fluidic-material consists of a 70 × 70 array of meta-molecules with a period of 300 μm, as shown in Fig. 7.7a. The liquid metal is injected into the bottom PDMS layer with 10-μm vent holes and sealed afterward. The heights of the liquid–metal pillars are controlled by using a syringe pump connected to the input of the top PDMS layer. The pumping speed of the syringe pump is set to be 3.6 mL/min to avoid an overwhelming instantaneous pressure within the top and bottom PDMS layers. As a result, the tuning speed of the liquid–metal pillars is approximately 10 ms when their heights vary from 0 to 100 μm, including the stabilization time. The heights of the liquid–metal pillars are stabilized by the capillary effect of the silicon holes. The forces induced by liquid metal surface tension are approximately 100 μN, which can be estimated by the surface tension coefficient, the contact angle, and the radius of the liquid–metal pillar. The gravity of each liquid–metal pillar is approximately 0.1 μN, which is three orders of magnitude smaller than the force induced by surface tension. Therefore, the liquid metal is stabilized within the silicon holes when the pumping pressure is fixed.

The microscope images of the meta-molecules are shown in Fig. 7.7b, c without and with the liquid metal, respectively. The microscope is focused on the top surface of the silicon layer during the observation. The liquid metal has higher reflectivity than the PDMS and silicon, which is brighter in the graphs. The blur cross-shaped shadow is induced by the air-pumping microchannels within the top PDMS layer.

The experimental characterization of the meta-liquid-material is carried out by using a terahertz spectrometer (TPS Spectra 3000). Here, the THz beam is incident

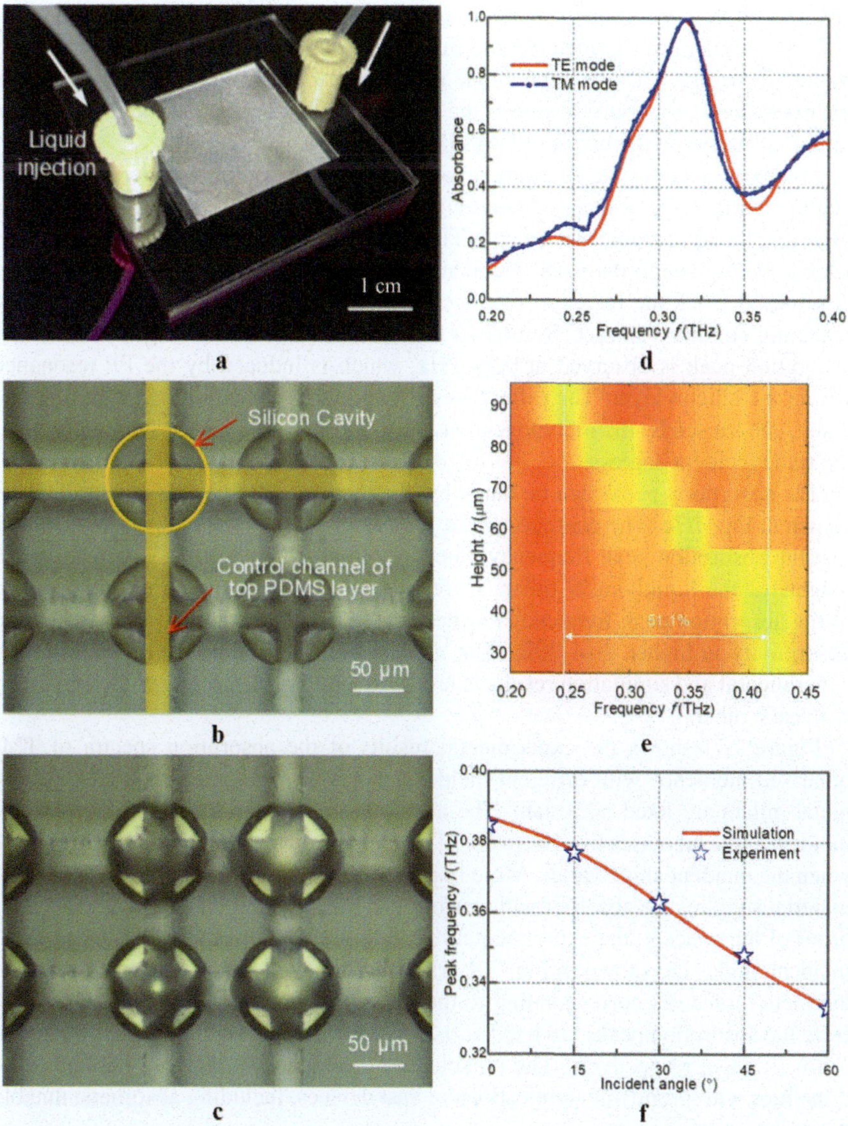

Fig. 7.7 **a** Photograph of the meta-fluidic-material. **b** and **c** Microscope image of a meta-molecule without and with liquid–metal injection, respectively. **d** Experimental results of the absorption spectra of the meta-fluidic-material with TE-polarized (red line) and TM-polarized (blue-dotted line) incidences when the heights of the liquid–metal pillars are $h = 70$ μm. **e** Absorption contour map of the tunable absorption spectra with normal incidence when h is tuned from $h = 30$ μm to $h = 90$ μm. **f** Measured absorption peak frequency as a function of incident angles at TM polarization state when $h = 50$ μm. A red-solid line and blue-pentagram symbols represent the simulation and experimental results

on the sample with a small incident angle (15°) tuned by a rotating mirror. The reflection spectra of the meta-fluidic-material are collected by a bolometer cooled by liquid Nitrogen. The incident angle and the reflection angle are identical during the experiment. The transmission of the meta-liquid-material is negligible due to the liquid metal within the bottom PDMS layer.

Figure 7.7d shows the absorption spectra of the meta-fluidic-material when the heights of the liquid–metal pillars are fixed at 70 μm. The red and blue-dotted lines represent the absorption spectra with TE and TM-polarized incidence, respectively, which overlap due to the small incident angle of 15°. The near-to-unity absorption is achieved when the incident frequency is 0.32 THz when the optical impedance matching condition is met. Similar to the simulation results in Fig. 7.6a, another absorption peak is observed at 0.29 THz, which is induced by the FP resonance within the silicon layer. The differences between the simulation and experimental results are caused by the inhomogeneous distribution of the liquid–metal pillars and the tilted incident angles.

The absorption peaks can be tuned by the heights of the liquid–metal pillars, as shown in Fig. 7.7e. The contour map of the meta-fluidic-material absorption shows that the absorption peak frequency blue shifts from 0.246 to 0.415 THz when the heights of the liquid–metal-pillars increase from 30 to 90 μm. The tuning range of the absorption peak frequency reaches 51.1% of the central frequency when the absorptivity is higher than 90%. The mismatch in the tuning range between the experimental and simulation results is less than 2%, which shows a good agreement with each other.

Figure 7.7f shows the experimental results of the absorption spectra of TM-polarized incidence with different incident angles when the heights of the liquid–metal pillars are fixed at 50 μm. The absorption peak slightly redshifts when the incident angle increases from 0° to 60°. The absorption peaks maintain at nearly 99% when the incident angle varies. More importantly, the tunability of the meta-fluidic-material supplies the absorber with performances adapted to the incident angles. As proof of the concept, the experimental results of fixed absorption peaks at different incident angles are shown in Fig. 7.8, which are realized by the adaptive control of the meta-molecules corresponding to the incident angles ranging from 15° to 60°. Here the absorption peaks are fixed at 0.25 THz, 0.28 THz, and 0.33 THz, as shown in Fig. 7.8a–c, respectively. The tunable and adaptable MFM offers various functionalities with promising applications in vast devices, including absorbers, tunable filters, beam steering, etc.

7.4 Summary

This chapter focuses on discussing MFM for near-unity absorption in structure designs, theoretical analysis, fabrications, and experiments. The design of the meta-fluidic-material is based on the impedance matching condition at the interface between the artificial material and its surrounding media, e.g., air. The near-unity

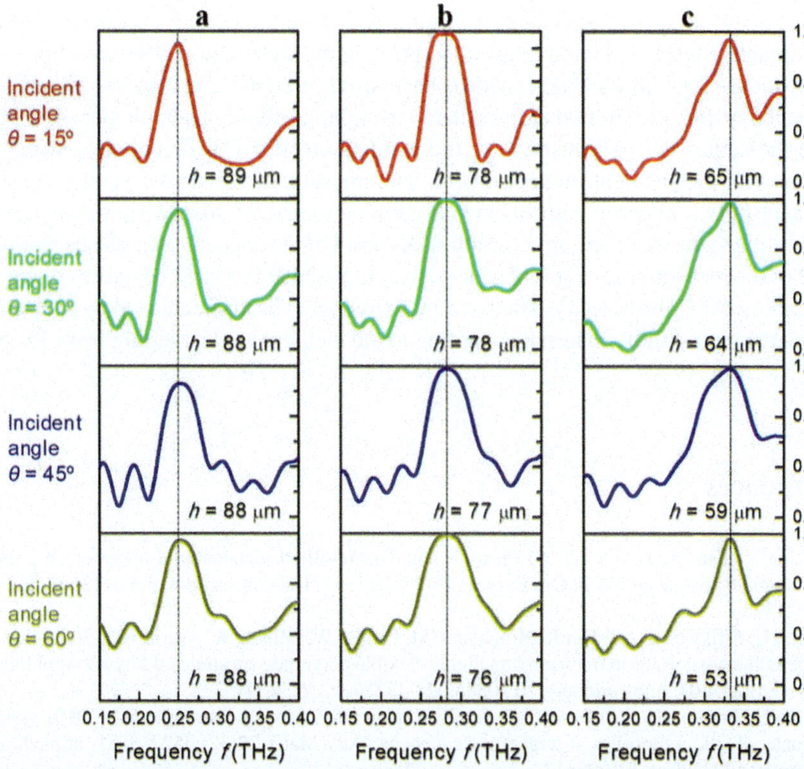

Fig. 7.8 Experimental results of the broadband and wide-angle absorber based on meta-fluidic-material. The incident THz waves are TM-polarized with incident angles of 15° (red lines), 30° (green lines), 45° (blue lines), and 60° (dark-yellow lines). **a**, **b** and **c** Show that absorption peaks can be fixed at a specific incident frequency by controlling the meta-molecules. As proof of the concept, the absorption peaks are fixed at 0.25 THz, 0.28 THz, and 0.33 THz, with the tuning range of the central frequency reaching 27.6%

absorption is achieved by rational designed electromagnetic resonances, which confine the incident electromagnetic materials within the lossy dielectric layers.

The water resonators induce strong magnetic resonances in the radiofrequency region, which can be designed to achieve the impedance matching condition in a broad bandwidth across the entire Ku, K, and Ka bands. In addition, microfluidic technology is applied to form and regulate the liquid-state resonators using PDMS microchannels. The demonstrated meta-fluidic-material absorber based on water resonators has a bandwidth of 78.9% central frequency with the above 90% absorptivity under normal incidence. More importantly, the high absorption and large bandwidth remain under oblique incidence with an incident angle ranging from 0° to 45°. In addition, an omnidirectional and broadband absorber is applied to a curved surface and tuned by injecting different concentrations of ethanol solution.

The meta-fluidic-material based on liquid–metal meta-molecules is demonstrated for frequency-agile and wide-angle absorption in the THz region. The liquid-pillar-based meta-molecules are designed to have strong magnetic resonances based on U-shaped resonators formed and regulated using microfluidic technologies. The 3D meta-molecules not only respond to the incident electrical fields but also strongly couple with the incident magnetic field. The microfluidic control system supplies the meta-fluidic-material with an extensive tuning range and adaptive performances, responding to the incident angle variation. As proof of the concept, a broad absorption band of the meta-fluidic-material is demonstrated, which is 51.1% central frequency with above 90% absorptivity. The fixed peak frequency is also achieved by adaptive tuning the meta-fluidic-material according to the incident angle, ranging from 15° to 60°.

References

1. Ling K, Kim K, Lim S (2015) Flexible liquid metal-filled metamaterial absorber on polydimethylsiloxane (PDMS). Opt Express 23(16):21375–21383. https://doi.org/10.1364/OE.23.021375
2. Tao H, Strikwerda AC, Fan K, Bingham CM, Padilla WJ, Zhang X, Averitt RD (2008) Terahertz metamaterials on free-standing highly-flexible polyimide substrates. J Phys D Appl Phys 41(23):232004. https://doi.org/10.1088/0022-3727/41/23/232004
3. Tao H, Amsden JJ, Strikwerda AC, Fan K, Kaplan DL, Zhang X, Omenetto FG (2010) Metamaterial silk composites at terahertz frequencies. Adv Mater 22(32):3527–3531. https://doi.org/10.1002/adma.201000412
4. Tao H, Chieffo LR, Brenckle MA, Siebert SM, Liu M, Strikwerda AC, Omenetto FG (2011) Metamaterials on paper as a sensing platform. Adv Mater 23(28):3197–3201. https://doi.org/10.1002/adma.201100163
5. Melik R, Unal E, Kosku Perkgoz N, Puttlitz C, Demir HV (2009) Flexible metamaterials for wireless strain sensing. Appl Phys Lett 95(18):181105. https://doi.org/10.1063/1.3250175
6. Iwaszczuk K, Strikwerda AC, Fan K, Zhang X, Averitt RD, Jepsen PU (2012) Flexible metamaterial absorbers for stealth applications at terahertz frequencies. Opt Express 20(1):635–643. https://doi.org/10.1364/OE.20.000635
7. Xu X, Peng B, Li D, Zhang J, Wong LM, Zhang Q, Xiong Q (2011) Flexible visible–infrared metamaterials and their applications in highly sensitive chemical and biological sensing. Nano Lett 11(8):3232–3238. https://doi.org/10.1021/nl2014982
8. Pryce IM, Aydin K, Kelaita YA, Briggs RM, Atwater HA (2010) Highly strained compliant optical metamaterials with large frequency tunability. Nano Lett 10(10):4222–4227. https://doi.org/10.1021/nl102684x
9. Li H, Wang G-M, Hu G, Cai T, Qiu C-W, Xu H-X (2020) 3D-printed curved metasurface with multifunctional wavefronts. Adv Opt Mater 8(15):2000129. https://doi.org/10.1002/adom.202000129
10. Chanda D, Shigeta K, Gupta S, Cain T, Carlson A, Mihi A, Rogers JA (2011) Large-area flexible 3D optical negative index metamaterial formed by nanotransfer printing. Nat Nanotechnol 6(7):402–407. https://doi.org/10.1038/nnano.2011.82
11. Yoo YJ, Zheng HY, Kim YJ, Rhee JY, Kang JH, Kim KW, Lee YP (2014) Flexible and elastic metamaterial absorber for low frequency, based on small-size unit cell. Appl Phys Lett 105(4):041902. https://doi.org/10.1063/1.4885095

12. Wang X, Chen J, Guo T, Shi Y (2020) Polarization tunable color filters based on all-dielectric metasurfaces on a flexible substrate. Opt Express 28(15):21704–21712. https://doi.org/10.1364/OE.398494
13. Kasirga TS, Ertas YN, Bayindir M (2009) Microfluidics for reconfigurable electromagnetic metamaterials. Appl Phys Lett 95(21):214102. https://doi.org/10.1063/1.3268448
14. Wang J, Liu S, Guruswamy S, Nahata A (2013) Reconfigurable liquid metal based terahertz metamaterials via selective erasure and refilling to the unit cell level. Appl Phys Lett 103(22):221116. https://doi.org/10.1063/1.4837675
15. Andryieuski A, Kuznetsova SM, Zhukovsky SV, Kivshar YS, Lavrinenko AV (2015) Water: Promising opportunities for tunable all-dielectric electromagnetic metamaterials. Sci Rep 5(1):13535. https://doi.org/10.1038/srep13535
16. Rybin MV, Filonov DS, Samusev KB, Belov PA, Kivshar YS, Limonov MF (2015) Phase diagram for the transition from photonic crystals to dielectric metamaterials. Nat Commun 6(1):10102. https://doi.org/10.1038/ncomms10102
17. Yoo YJ, Ju S, Park SY, Ju Kim Y, Bong J, Lim T, Lee Y (2015) Metamaterial absorber for electromagnetic waves in periodic water droplets. Sci Rep 5(1):14018. https://doi.org/10.1038/srep14018
18. Zhu WM, Dong B, Song QH, Zhang W, Huang RF, Ting SK, Liu AQ (2014, 26–30 Jan) Tunable meta-fluidic-materials base on multilayered microfluidic system. Paper presented at the 2014 IEEE 27th international conference on micro electro mechanical systems (MEMS)
19. Chen X, Grzegorczyk TM, Wu B-I, Pacheco J, Kong JA (2004) Robust method to retrieve the constitutive effective parameters of metamaterials. Phys Rev E 70(1):016608. https://doi.org/10.1103/PhysRevE.70.016608

Chapter 8
Adaptive Metasurfaces for Dispersion Control

8.1 Introduction

The word "adaptive optics" was first coined by Horace W. Babcock in 1953 and then referenced by Poul Anderson in his novel "Tau Zero" in 1970 [1]. Two decades later, adaptive optics (AO) technology was applied to optical systems for improved performances by correcting the incoming wavefront distortions using deformable mirrors. Now the AO technologies have been widely used in laser communications [2, 3], astronomical telescopes [4, 5], advanced microscopic [6, 7], and imaging systems [8, 9] for atmospheric distortion control. The incident wavefronts can be measured and controlled by AOs with deformable mirrors or liquid–crystal spatial light modulators. Interestingly, MEMS technology has long been applied to AO devices where deformable mirrors or phase shifters are used in wavefront shaping applications with a microscale spatial resolution limited by state-of-art fabrication technologies.

On the other hand, the developments of micro-optics based on metasurfaces supply the AO devices with fast tuning speed and compact sizes [10, 11]. More importantly, the new physics of MEMS and microfluidic metasurfaces enable improved performances and new functionalities of electromagnetic devices with controllable and reconfigurable wavefronts. Therefore, the tunable and reconfigurable metasurfaces are promising candidates for adaptive functionalities, accommodating non-perfect inputs, e.g., variation of incident angles, frequencies, wavefront deformations, etc., which is named adaptive metasurfaces.

The adaptive meta-fluidic-material for the variations of the incident angle is discussed in Chap. 7, which functions as a broadband absorber. In this chapter, adaptive metasurfaces are designed to accommodate the dispersion of meta-molecules induced by their resonance features, which is essential for electromagnetic devices when broad working bandwidths are needed. The dispersions of metasurface devices are caused by both the resonance features of the meta-molecules and propagation-phase differences of the electromagnetic waves with different frequencies. For example, a metasurface flat lens reconfigures the incident wavefronts into concaved

W. Zhu and A.-Q. Liu, *Metasurfaces: Towards Tunable and Reconfigurable Meta-devices*, Microfluidics and Nanophotonics: Science and Engineering 1,
https://doi.org/10.1007/978-981-19-6925-6_8

ones by meta-molecules spatially arranged within a plane. The meta-molecule may induce various phase shifts to incidences with different frequencies, which results in a dispersion of the output spatial phase distributions. On the other hand, distances between spatially arranged meta-molecules and the focal point are different, which results in the dispersion of the focal point of the flat lens. Recently, many dispersionless metasurface lenses have been demonstrated, some of which even suppress the overall dispersion of the metasurface lens by the above-mentioned two effects. Nevertheless, the dispersionless metasurface devices are either working on discreet frequencies or limited bandwidths [12].

In the meantime, tunable and reconfigurable metasurfaces with controllable phase gradients are good candidates to control the dispersion adaptively. The dispersion can be suppressed by reconfiguring the metasurface to accommodate electromagnetic waves with various frequencies, leading to frequency-independent functionalities. This chapter focuses on the metasurfaces which apply liquid–metal meta-molecules to design the adaptive metasurface for dispersion control. The geometries of the liquid–metal meta-molecules can be reconfigured to impart desired phase profiles when the incident frequency varies, which is enabled by using the microfluidic control systems. The microfluidic metasurface is adaptive to electromagnetic waves with different frequencies to achieve the same functionality within a broadband incident frequency range [13, 14].

8.2 Adaptive Metasurfaces with Dynamic Dispersion Control

The Laws of reflection and refraction of a homogeneous interface can be derived from Fermat's principle as [15],

$$\begin{cases} n_t \cdot \sin\theta_t - n_i \cdot \sin\theta_i = 0 \\ n_r \cdot \sin\theta_r - n_i \cdot \sin\theta_i = 0 \end{cases} \tag{8.1}$$

where n_t, n_r, and n_i are the refractive indices of the media for refracted, reflected, and incident electromagnetic waves, respectively. And θ_t, θ_r, and θ_i are the refraction, reflection, and incidence angles, respectively. In most cases, the incident and reflected electromagnetic waves propagate within the same media, resulting in the equivalence of the incident and reflection angles $\theta_i = \theta_r$. However, the Laws of reflection and refraction only hold when the electromagnetic waves are incident on a homogeneous interface where the phase shifts ϕ for the refracted or reflected electromagnetic waves are constant. In other words, the gradient of the phase shift is equal to zero $d\phi/dx = 0$, considering an interface along the x-direction, which is not correct for nonhomogeneous interfaces, e.g., metasurfaces. The meta-molecules determine the phase shift of the metasurfaces for a specific incidence. As a result, the phase shift can be written as $\phi(x)$, which is not a constant for nonhomogeneous metasurfaces.

As shown below, Professor Capasso uses the generalized Snell's Law to describe the refraction and reflection of the electromagnetic waves incident on nonhomogeneous interfaces.

$$\begin{cases} n_t \cdot \sin\theta_t - n_i \cdot \sin\theta_i = \dfrac{\lambda_0}{2\pi} \cdot \dfrac{d\phi}{dx} \\ n_r \cdot \sin\theta_r - n_i \cdot \sin\theta_i = \dfrac{\lambda_0}{2\pi} \cdot \dfrac{d\phi}{dx} \end{cases} \tag{8.2}$$

where λ_0 is the incident wavelength in the free space. Equation 8.1 can also be written by using wave numbers as,

$$\begin{cases} k_t \cdot \sin\theta_t - k_i \cdot \sin\theta_i = \dfrac{d\phi}{dx} \\ k_r \cdot \sin\theta_r - k_i \cdot \sin\theta_i = \dfrac{d\phi}{dx} \end{cases} \tag{8.3}$$

where $k_i = n_i \cdot 2\pi/\lambda_0, k_r = n_r \cdot 2\pi/\lambda_0$, and $k_t = n_t \cdot 2\pi/\lambda_0$ are the wavenumbers of the incident, reflected and refracted electromagnetic waves, respectively. Equation 8.3 can be derived by using the momentum conservation at the interface, where the phase gradient induces an extra momentum $d\phi/dx$. Therefore, refraction angle θ_t and reflection angle θ_r are functions of both incident wavelength λ and phase gradient.

Figure 8.1a shows the schematic of a reflective metasurface, which is designed by using spatially arranged meta-molecules forming periodical supercells with effective lengths ξ. Most metasurface devices have supercell structures for two reasons: The phase shift ϕ cannot go infinite by changing the meta-molecule's design. Therefore, ϕ is folded every 2π intervals. The other reason is that the supercell structure can reduce the design difficulties of the meta-molecules by selecting only a group of meta-molecules, as shown in Fig. 8.1e. As a result, the reflection angle of the incidence is determined by both the phased gradient as in Eq. 8.3 and the diffraction effect described below,

$$\begin{cases} \theta_r = \sin^{-1}\left(\dfrac{c}{\xi \cdot f_i}\right) \\ \theta_r = \sin^{-1}\left(\dfrac{d\phi}{dx \cdot k_r}\right) \end{cases} \tag{8.4}$$

where c and f_i are the velocity of the electromagnetic waves in vacuum and the incident frequency, respectively. The incident electromagnetic waves have an incident angle $\theta_i = 0$, as shown in Fig. 8.1a. Equation 8.4 shows that the reflection angle can be fixed by changing the phase gradient and the effective supercell length ξ when the incident frequency is changed.

Another example is the lens dispersion, as shown in Fig. 8.1b. The traditional lenses have focal length f, which can be derived by the Lens-maker's equation as shown below,

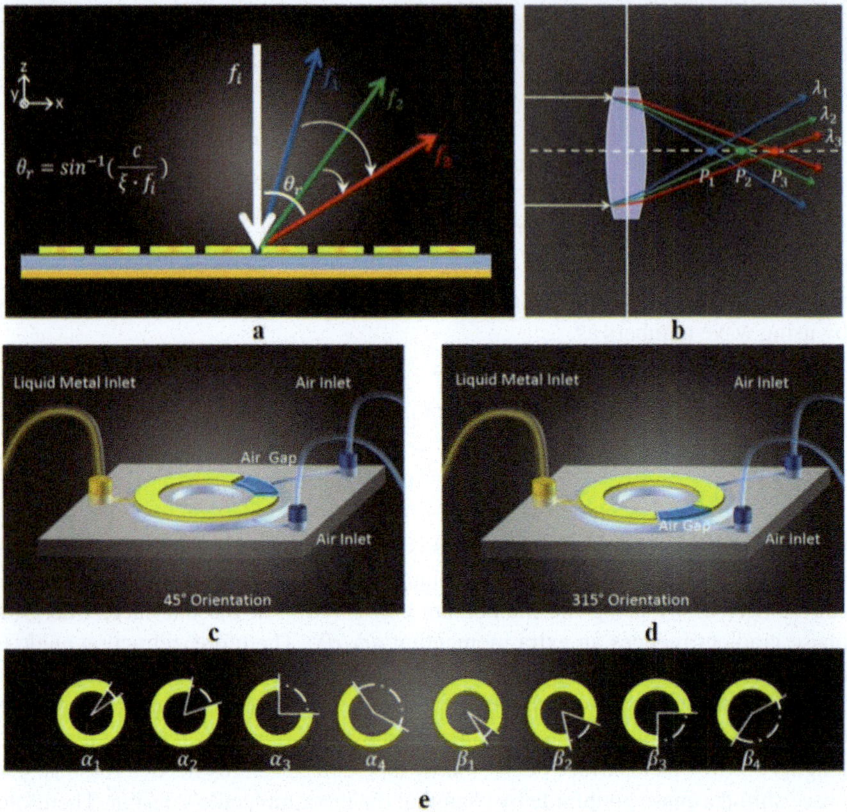

Fig. 8.1 Schematics of **a** dispersive grating and **b** lens; The control of the meta-molecules with the air-gap orientation of **c** 45° and **d** 315°; **e** The supercell of the adaptive metasurface

$$f = \frac{1}{(n-1)\left[\frac{1}{R_1} - \frac{1}{R_2} + \frac{(n-1)T}{nR_1R_2}\right]} \tag{8.5}$$

where n is the refractive index of the lens compositing materials, and the surrounding media is assumed to be the vacuum. R_1 and R_2 are the front and back radius of the lens, respectively, and T is the thickness of the lens. The traditional lenses have bulky sizes which can converge light into focal distances that vary with the incident wavelength, as shown in Fig. 8.1b. The dispersions of refractive lenses are mainly due to their constituent materials, the refractive indexes of which are functions of incident wavelength $n(\lambda)$. As a result, the focal length f also becomes a function of the incident wavelength.

On the other hand, metasurfaces are ultra-thin 2D materials that can tailor the wavefront of the light. It is possible to design a flat lens based on metasurface, much thinner than traditional lenses. However, the focal length varies when the incident frequency is changed due to the dispersive nature of meta-molecules. The design of a metasurface lens is different from the traditional lens based on the Lens-maker equation. Every subwavelength meta-molecules function as phase shifters with a designed spatial phase gradient when a plane wave with wavelength λ incidents on the metasurface. The wavelets of each meta-molecule can be approximately regarded as spherical waves when their sizes are much smaller than the wavelength. The sum of the scattered wavelets forms the overall refracted and reflected wavefront. The phase distribution of a hyperbolic metasurface lens can be written as

$$\phi(x) = \frac{2\pi}{\lambda_0}\left(\sqrt{x^2 + f^2} - f\right) \tag{8.6}$$

where x represents the spatial distributions of the meta-molecules. The phase shift $\phi(x)$ has to be changed when the incident frequency varies to fix the focal length f. Therefore, one can achieve the same focal length for different incident frequencies by controlling the phase shift imparted by the metasurface.

The adaptive metasurface comprises liquid–metal meta-molecules designed with reconfigurable geometry (shape and orientation), as shown in Fig. 8.1c, d. The yellow part represents the liquid metal. Specifically, the reconfiguration of the meta-molecule is realized by using microfluidic control systems. The meta-molecules consist of a liquid metal inlet and two air inlets to control the shapes of the liquid metal, whereby its shape and orientation are reconfigurable, as shown in Fig. 8.1c, d. In the meantime, Fig. 8.1e shows the schematic illustration of a supercell with eight ring-shaped meta-molecules where a 2π phase shift is achieved by reconfiguring the shapes and orientations of the liquid–metal meta-molecules. The microfluidic metasurfaces can be formed and controlled by the following processes: Firstly, liquid metal is pumped into the microfluidic channels and forms ring-shaped meta-molecules. Then, air bubbles are generated within the liquid metal when air is injected into the microchannels, forming the meta-molecules' gaps. Here, the air gaps can be tuned continuously in the range between 5° and 180° by changing the input pressure of the air. The shapes of the meta-molecules determine the phase shifts of the cross-polarized transmission. The phase shift from 0 to π can be achieved by changing the gap sizes only with a fixed orientation of the meta-molecules. However, it is essential to reconfigure the orientations of the meta-molecules to obtain a 2π phase modulation. The orientation reconfigurations are achieved by selectively pumping air into one of the two air inlets, as shown in Fig. 8.1c, d. The orientation of the meta-molecules is designed to be selectively reconfigured into two perpendicular directions, i.e., 45° and 315°. A supercell of the reconfigurable metasurface is formed by periodical variation of the meta-molecules' geometries, as shown in Fig. 8.1e.

8.3 A Flat Lens Based on Adaptive Metasurfaces

Figure 8.2a shows a metasurface flat lens with a fixed focal length, which is adaptively tuned according to the incident frequency. The flat lens is based on an adaptive metasurface composed of a layer of microfluidic metasurface sandwiched by two orthogonal metal grating layers. The microfluidic system can control the meta-molecules in gap openings and orientations, as shown in Fig. 8.2b. The tuning mechanisms have been discussed in the previous section, and one can find the design and fabrication processes in Chap. 3, which will not be further explained in this section. All the split-ring-shaped microchannels for the meta-molecules are the same at the middle layer. Therefore, the inner and outer radius of all the meta-molecules is identical. The gap opening of the meta-molecules is defined by the arc angle of the air gap G, as shown in Fig. 8.1b, c. The phase gradient of the adaptive metasurface is realized by the spatial variations of the meta-molecules' gap openings and orientations. Here, the adaptive metasurface has a square array of meta-molecules. Only electric dipole resonance can be excited within the ring-shaped meta-molecules without the gap opening, i.e., $G = 0°$. As a result, the meta-molecule has an absorption resonance at the wavelength proportion to the half-length of the liquid–metal ring. On the other hand, this resonance can be gradually tuned by removing a section of the liquid–metal ring when the air bubble is formed and tuned within the microchannel filled with the liquid metal. As a result, the lens function of the adaptive metasurface can be realized and tuned by the reconfiguration of the meta-molecules' spatial distributions, i.e., phase shift distributions $\phi(x)$.

Figure 8.2c, d show the amplitude and phase contour maps of cross-polarized transmission as functions of the frequency and gap opening when the orientation of the meta-molecule is 45°. The periodical boundary condition is used in the simulation while the incidence is linear polarized. One can reach the π phase shift by changing the gap opening of the meta-molecules from 10° to 200° with a fixed orientation angle when the incident frequency ranges from 10 to 20 GHz, as shown in Fig. 8.2d. As a result, the phase shift covers 2π with the gap opening variation when the orientation angle switches between 45° and 315°. The normalized amplitudes of the meta-molecules are shown in Fig. 8.2c, which shows that the amplitude variation is relatively small when the incident frequency varies from 10 to 20 GHz. The transmission phase can be tuned up to 2π with the gap opening ranging from 10° to 180° range while maintaining the normalized amplitude larger than 0.7 when the incident frequency is 13 GHz and 14 GHz. The tuning range of the gap opening has to be 150° for a 2-π phase shift when the incident frequency is 15, 16, and 17 GHz. Therefore, it is feasible to control the dispersion of the flat lens by adaptive tailoring the output wavefront according to the incident frequency.

The far-field intensity distributions in the xz-plane for different metasurface configurations are shown in Fig. 8.3a–e. Here, the metalens is placed on the xy-plane while the incident electromagnetic waves propagate along the z-direction. As a proof of the concept, the lens function is demonstrated by the intensity distributions of the output from the metasurface adaptively tuned to accommodate five different

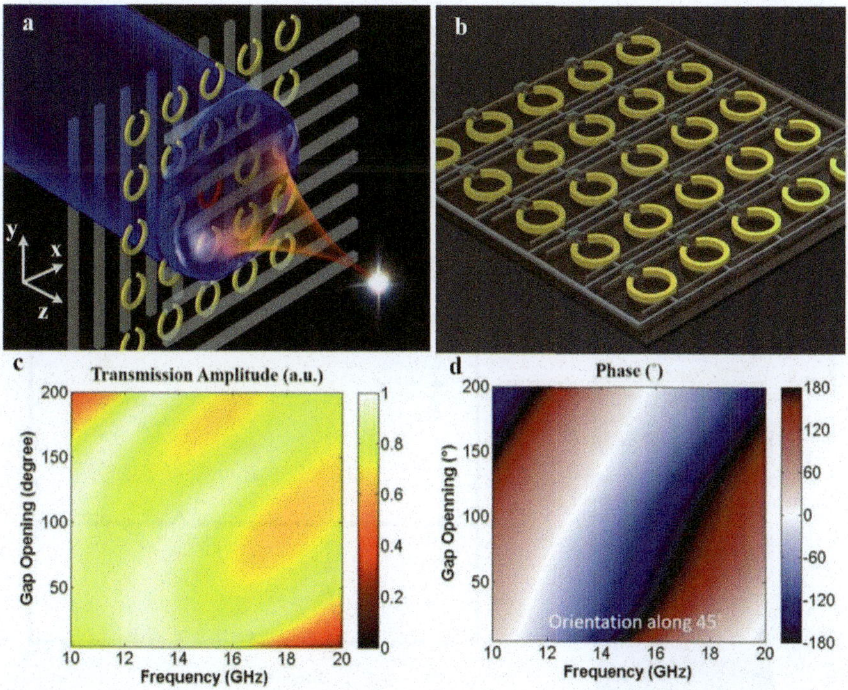

Fig. 8.2 The schematics of the adaptive metasurface and the simulation results of the meta-molecules. The schematic of **a** the adaptive metasurface. **b** The middle layer of the reconfigurable meta-molecules. The liquid metal is controlled by the air pumped from the microchannels, which are much smaller than the microchannels filled with liquid metal. Amplitude (**c**) and phase (**d**) of the meta-molecules as functions of the frequency and gap opening when the gap orientation is 45°

incident frequencies, which are 13 GHz (Fig. 8.3a), 14 GHz (Fig. 8.3b), 15 GHz (Fig. 8.3c), 16 GHz (Fig. 8.3d) and 17 GHz (Fig. 8.3e). The focal lengths of the metasurface flat lens are fixed at 200 mm when the incident frequency varies. Figure 8.3f shows the focal lengths of the reconfigured metasurface lens as functions of the incident frequency. The symbols represent the metasurface flat lens with different configurations. The metasurface flat lens is dispersive at any particular configuration. For example, the focal lengths of one metasurface configuration denoted as red stars vary from 200 to 287 mm when the incident frequency increases from 13 to 17 GHz. However, different configurations of metasurfaces result in a focal length of 200 mm, represented by the black line, with varying frequencies of the incident, which makes it possible to fix the focal length by changing the configuration metasurface to accommodate the incident frequency. In other words, the concept of the adaptive metasurface is validated.

In fabrication, soft lithography is used to fabricate the reconfigurable metasurface, as discussed in Chap. 3. Here, a binary pneumatic microfluidic control system is designed to control the geometries of the meta-molecules individually. One control

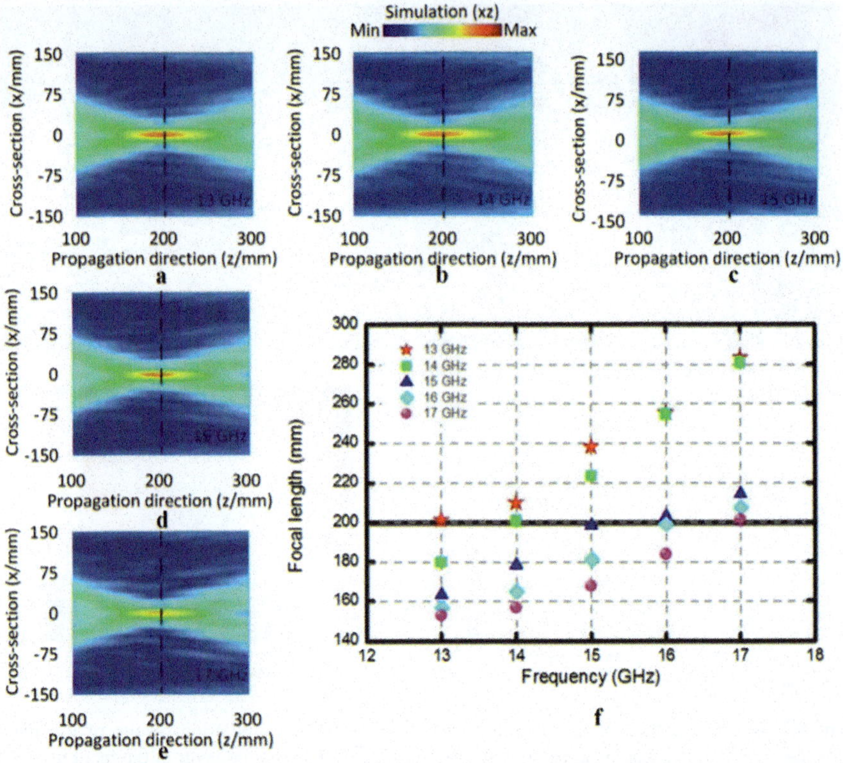

Fig. 8.3 Simulation results of the metalens based on the adaptive metasurface. **a–e** The electromagnetic wave intensity distributions of metalens, showing the focal length at 200 mm when the metasurface is adaptively tuned according to the incident frequency ranging from 13 to 17 GHz. **f** The flat lens' focal length as a function of the incident frequencies. The symbols represent different metasurface configurations

unit consists of two microchannels, defined as one bit. Similar to the digital circuit widely used in display, a 10-bit binary control system can control at most 210 = 1024 air channels individually connected to the meta-molecules. The ring-shaped microfluidic channels of the meta-molecules are imprinted into a 2-mm thick PDMS layer. The PDMS layer is bonded to a 1-mm thick solid PMMA layer to avoid the deformation of the microfluidic channels.

The adaptive metasurface consists of 60 × 60 meta-molecules with a period of 5 mm and a total footprint of o 300 mm × 300 mm, as shown in Fig. 8.4a. The adaptive metasurface comprises nine arrays of 20 × 20 meta-molecules independently controlled by nine microfluidic binary control systems. The inserts show an individual tuning of one meta-molecule's gap-opening, highlighted by a false yellow color. The hydrochloric vapor is pumped into the air channel connected to the meta-molecule, expels the liquid metal (galinstan) away, and forms a gap-opening of 120°.

Then the gap-opening restores to 5° when the pumping pressure is reduced. Here, the hydrochloric vapor removes the oxidized galinstan during the tuning process.

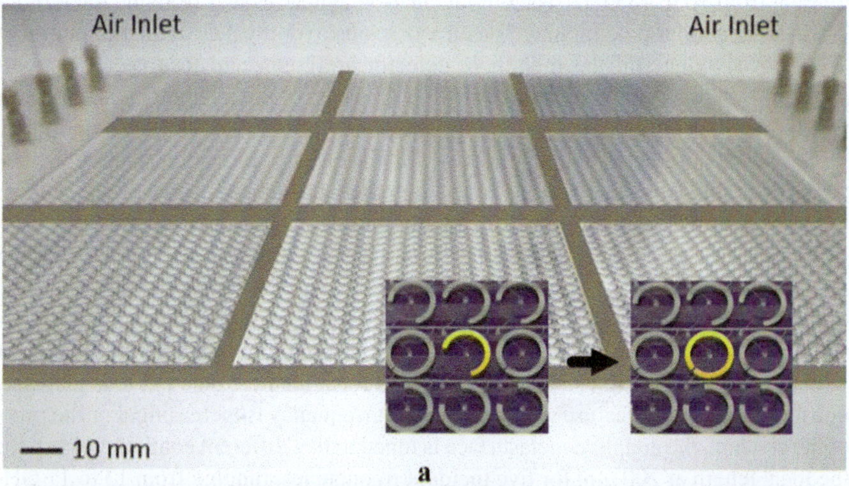

a

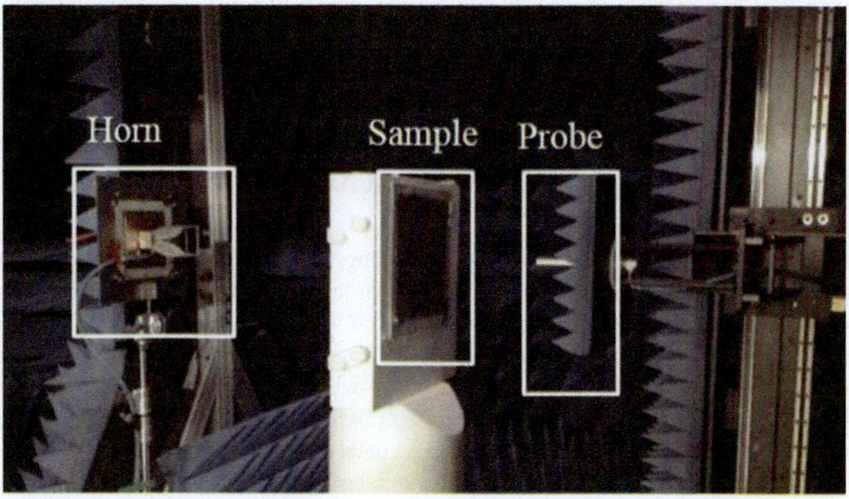

b

Fig. 8.4 Photographs of the adaptive metasurface and the experimental setup. **a** An overview photograph of the adaptive metasurface. The inserts show zoomed-in views of a single meta-molecule with a gap-opening of 120°, which is formed by pumping the liquid metal into a ring-shaped microchannel. The gap-opening reduces to 5° when the input pressure of hydrochloric vapor decreases. The liquid metal within the reconfigured meta-molecule is highlighted with false color. **b** Measurement setup for near-field measurement of the metasurface flat lens

Figure 8.4b shows the experimental setup of the flat lens based on the adaptive metasurface, which is located in the microwave anechoic chamber. The incident electromagnetic wave is from a horn antenna, radiating a Gaussian wave with frequencies ranging from 10 to 18 GHz. The output electromagnetic waves from the flat lens are detected by a monopole mounted on an xyz scanner for the near-field measurement. The horn antenna and the monopole detector are connected to a vector network analyzer (VNA). The flat lens is located on the waist of the incident Gaussian beam for all the incident frequencies to mimic the plane wave incidence. The monopole is set to receive the electromagnetic waves with the cross-polarization state of the incidence. The intensity distributions of the output electromagnetic waves are mapped in xy-plane at different z distances from the sample with a step of 5 mm.

Figure 8.5 shows the experimental results of the flat lens based on the adaptive metasurface with different configurations. The flat lens is illuminated by a Gaussian beam from a horn antenna, propagating along the z-direction. The flat lens is placed at a xy-plane located at the waist of the Gaussian beam to mimic a plane wave incidence with a linear polarization state. The measured focal length of the flat lens increases due to the dispersion feature when the incident frequency is increasing. For the proof of the concept, the adaptive metasurface is tuned to five different configurations to fix the focal length at 200 mm for five incident frequencies, ranging from 13 to 17 GHz with one GHz in step. The experimental results agree well with the simulation results, as shown in Fig. 8.3a–e, proofing that the adaptive metasurface can be applied to the flat lens with dynamic dispersion control.

8.4 Anomalous Reflection Based on Adaptive Metasurface

A broadband reflection can be realized by using traditional geometry optics when the incident angle is equal to the reflected angle based on Snell's law. However, diffractive optics have limited bandwidth due to the dispersion, which is similar to the devices based on Anomalous reflection. The Anomalous reflection is brought forward based on the generalized Snell's law when the angle of reflection is not the same as that of the incidence due to the phase shift gradience of the interface, e.g., metasurfaces. Based on Eq. (8.4), two conditions must be satisfied to maintain the Anomalous reflection angle when the incident frequency varies. One is that the $1/(\xi \cdot f_i)$ must be a constant when f_i changes. In other words, the effective length of the supercell has to be tuned according to the incident frequency. The other condition is that the phase gradient $d\phi/dx$ must be tuned when the incident frequency changes. These two conditions can only be satisfied at discreet incident frequencies since the meta-molecule's period cannot be infinitely small, limiting the supercell's effective length choices.

In this section, the Anomalous reflection with a fixed angle is realized using an adaptive metasurface with three incident frequencies, which are 10.5, 12, and 14 GHz. The effective length of the supercell ξ varies from 40 to 30 mm when the incident frequency increases. The supercell configurations for 10.5 GHz, 12 GHz,

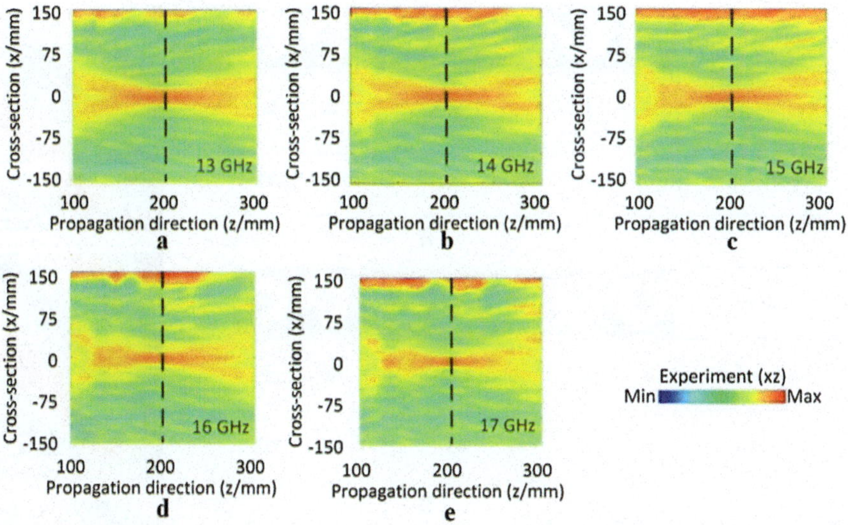

Fig. 8.5 Experimental results of the flat lens based on the adaptive metasurface. **a–e** Intensity distributions in the xz-plane of the metasurface flat lens's output when the metasurface is adaptively reconfigured to accommodate the incident frequency, ranging from 13 to 17 GHz. **f** Simulation (colored-solid lines) and experimental results (colored symbols) of reflection angle as a function of frequency from 10 to 15 GHz

and 14 GHz incidences are denoted as C10.5, C12, and C14, respectively, as shown in Fig. 8.6. The effective length of the supercell ξ is calculated by using Eq. 8.4 with fixed θ_r for different incident frequencies. The cross-polarized phase distributions of the meta-molecules' reflection are shown in the right column of Fig. 8.6. Here the numerical simulations are based on each meta-molecules with periodical boundary conditions. The initial phases of incident plane waves are identical for each simulation. At the same time, the frequencies of the incidence are 10.5 GHz, 12 GHz, and 14 GHz for different supercell configurations, as shown in Fig. 8.6a–c, respectively. The simulation results of eight adjacent meta-molecules for each configuration are grouped to show the reflected wavefront defined by the phase distribution's contour lines (white-solid lines). All the wavefronts and metasurface have the same separation angle (reflection angle), indicating a fixed angle of reflection.

Figure 8.7 shows the simulation results of the abnormal reflection from the metasurface with configuration C10 (Fig. 8.7a, b), C12 (Fig. 8.7c, d), and C14 (Fig. 8.7e, f), and the incident frequency is 10.5 GHz, 12 GHz, and 14 GHz, respectively. The left column shows the scattering field distributions, while the right column shows the far-field scattering intensity. The incident electromagnetic waves are propagating along the direction perpendicular to the metasurface. The anomalous reflection angles are fixed at $-45°$ for all the three configurations with different incident frequencies. The wavefront is smoother in configuration C10 than those of the C12 and C14 configurations since configuration C10 has eight orders of meta-molecules. There

Configuration($C_{10.5}$)	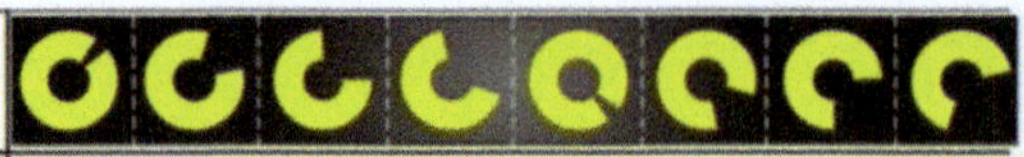							
Gap Opening (°)	5	45	95	130	5	45	95	130
Orientation (°)	45	45	45	45	315	315	315	315

a

Configuration(C_{12})								-
Gap Opening (°)	5	35	90	140	15	60	115	-
Orientation (°)	45	45	45	45	315	315	315	-

b

Configuration($C_{10.5}$)							-	-
Gap Opening (°)	5	45	105	5	45	105	-	-
Orientation (°)	45	45	45	315	315	315	-	-

c

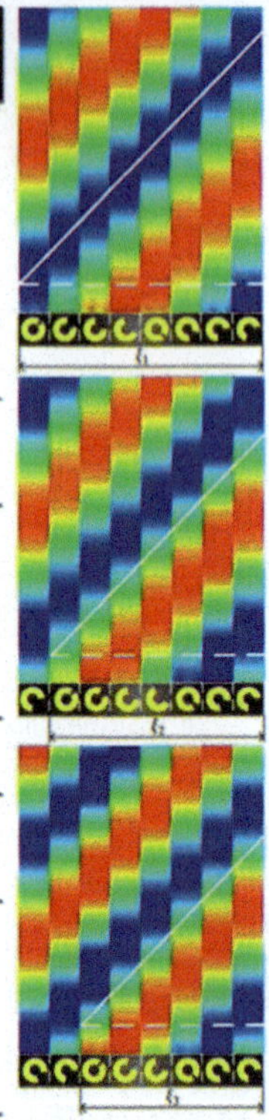

Fig. 8.6 Configurations of the adaptive metasurface for dispersionless beam steering. The contour maps show the simulation results of the electric fields scattered by the metasurface with different configurations, i.e., **a** C10.5, **b** C12, and **c** C14

are only seven orders and six orders of meta-molecules in configurations C12 and C14, respectively. Here, the orders of meta-molecules are defined by the numbers of the meta-molecules within one supercell, which represent the continuity of the spatial distribution of phase shifts with a fixed period. As a result, the wavefront discontinuity results from the phase discontinuity introduced by the limited orders of meta-molecules. The wavefront control can be improved by increasing the orders of meta-molecules within one supercell of the adaptive metasurface with a shorter period of meta-molecules.

Figure 8.7b, d, f show the cross-polarized scattering intensity distributions at different angles when the adaptive metasurface is tuned to C10.5, C12, and C14 configurations, respectively. The blue-solid lines represent the normalized simulation results, while the red-dashed lines represent the experimental results. In all configurations, the anomalous reflection angles are fixed at – 45°. In the meantime, the measured diffraction efficiency drops from 50.7% to 30.5% due to the loss induced by the phase discontinuities, which worsen as the operating frequency increases from

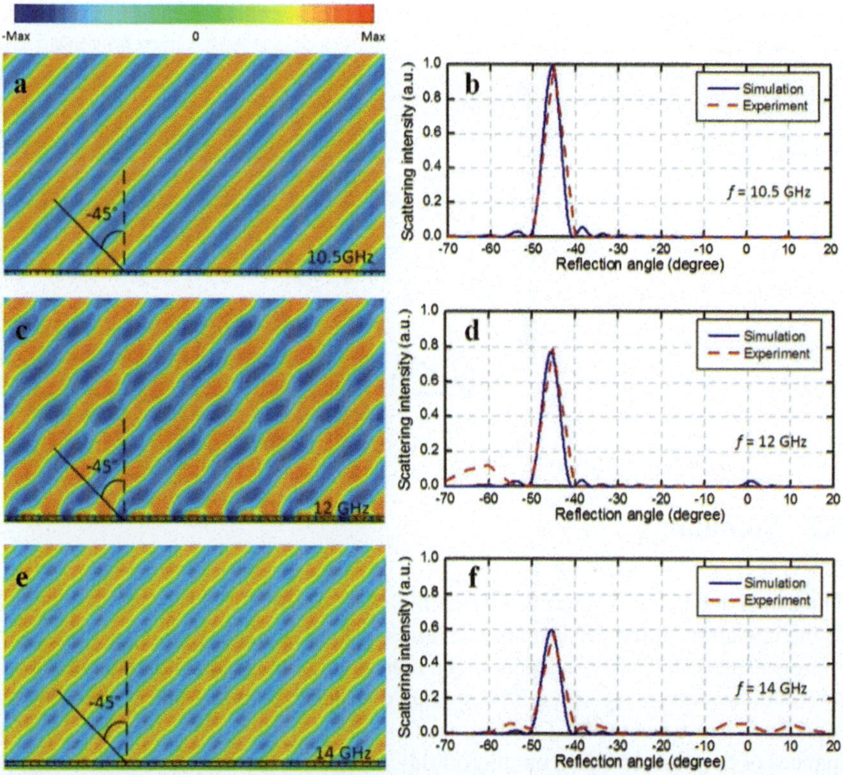

Fig. 8.7 Simulation and experimental results of the adaptive metasurface's E-field distribution and far-field scattering intensities at different configurations. The anomalous reflection angles are fixed at 45° under the normal incidence with the incident frequency of **a**, **b** 10.5 GHz, **c**, **d** 12 GHz, and **e**, **f** 14 GHz, respectively

10.5 to 14 GHz. The phase discontinuities can be improved by increasing the number of the meta-molecules within one supercell with a shorter period. The simulation and experimental results are in good agreement with each other.

Figure 8.8 shows the anomalous reflection angle as a function of incident frequencies when the adaptive metasurface is tuned to C10.5 (red-line and symbols), C12 (green-line and symbols), and C14 (blue-line and symbols) configurations, respectively. The colored lines represent the simulation results, while the symbols represent the experimental results. The anomalous reflection angle decreases with frequency at a fixed configuration due to the dispersion effect similar to a grating. Take configuration C10.5 as an example; the anomalous reflection angle changes from − 48.5° to − 30° when the incident frequency increases from 10 to 15 GHz. However, the anomalous reflection angle can be fixed at − 45° by tuning the adaptive metasurface to different configurations, as highlighted by the black line in Fig. 8.8.

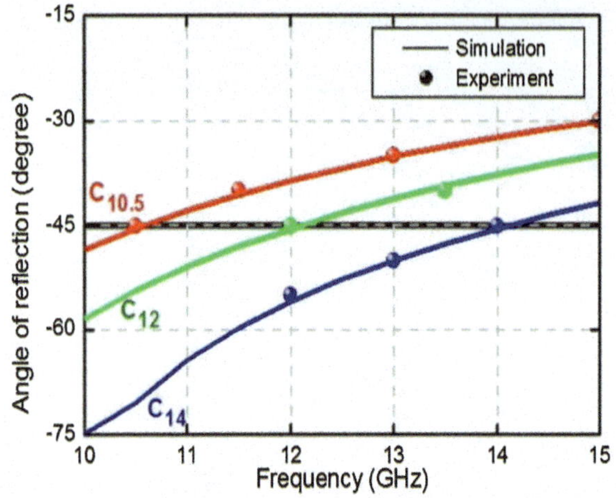

Fig. 8.8 Simulation results (colored lines) and experimental results (colored symbols) of reflection angle as a function of frequency from 10 to 15 GHz

8.5 Summary

The dispersive nature of metasurfaces limits their working bandwidths, which leads to the deviation of the performances as the working frequency changes. The adaptive metasurface consists of liquid–metal meta-molecules, which are regulated using microfluidic control systems. The spatial distribution of the liquid–metal meta-molecules can be controlled with great flexibility, enabling the adaptable performances of the devices based on microfluidic metasurfaces. As proof of the concept, the adaptive metasurface is designed to accommodate different incident frequencies to realize the same functionality with a dispersionless performance. The adaptive metasurfaces are demonstrated only in radiofrequency ranges due to the complexity of the microfluidic control system. It can be expected that the adaptive metasurfaces can be supplied with more functionalities and demonstrated at higher working frequencies with the development of microfluidic technology.

References

1. Anderson P, Brèque JD (2012) Tau Zéro: Le Bélial
2. Tyson RK (2002) Bit-error rate for free-space adaptive optics laser communications. J Opt Soc Am A 19(4):753–758. https://doi.org/10.1364/JOSAA.19.000753
3. Vorontsov M, Weyrauch T, Carhart G, Beresnev L (2010, 2010/02/03) Adaptive optics for free space laser communications. Paper presented at the lasers, sources and related photonic devices, San Diego, California
4. Sandler DG, Barrett TK, Palmer DA, Fugate RQ, Wild WJ (1991) Use of a neural network to control an adaptive optics system for an astronomical telescope. Nature 351(6324):300–302. https://doi.org/10.1038/351300a0
5. Ellerbroek BL (1994) First-order performance evaluation of adaptive-optics systems for atmospheric-turbulence compensation in extended-field-of-view astronomical telescopes. J Opt

Soc Am A 11(2):783–805. https://doi.org/10.1364/JOSAA.11.000783
6. Booth MJ (2007) Adaptive optics in microscopy. Philos Trans R Soc A: Math Phys Eng Sci 365(1861):2829–2843. https://doi.org/10.1098/rsta.2007.0013
7. Ji N (2017) Adaptive optical fluorescence microscopy. Nat Methods 14(4):374–380. https://doi.org/10.1038/nmeth.4218
8. Zhang J, Yang Q, Saito K, Nozato K, Williams DR, Rossi EA (2015) An adaptive optics imaging system designed for clinical use. Biomed Opt Express 6(6):2120–2137. https://doi.org/10.1364/BOE.6.002120
9. Mu Q, Cao Z, Hu L, Li D, Xuan L (2006) Adaptive optics imaging system based on a high-resolution liquid crystal on silicon device. Opt Express 14(18):8013–8018. https://doi.org/10.1364/OE.14.008013
10. Chen WT, Zhu AY, Capasso F (2020) Flat optics with dispersion-engineered metasurfaces. Nat Rev Mater 5(8):604–620. https://doi.org/10.1038/s41578-020-0203-3
11. Yao Y, Shankar R, Kats MA, Song Y, Kong J, Loncar M, Capasso F (2014) Electrically tunable metasurface perfect absorbers for ultrathin mid-infrared optical modulators. Nano Lett 14(11):6526–6532. https://doi.org/10.1021/nl503104n
12. Huang L, Chen X, Mühlenbernd H, Li G, Bai B, Tan Q, Zhang S (2012) Dispersionless phase discontinuities for controlling light propagation. Nano Lett 12(11):5750–5755. https://doi.org/10.1021/nl303031j
13. Zhu WM, Song QH, Yan LB, Zhang W, Wu PC, Chin LK, Liu AQ (2015, 21–25 June) Tunable flat lens based on microfluidic reconfigurable metasurface. Paper presented at the 2015 transducers—2015 18th international conference on solid-state sensors, actuators and microsystems (TRANSDUCERS)
14. Wu PC, Cai H, Gu YD, Zhu WM, Zhang W, Yang ZC, Liu AQ (2016, 24–28 Jan) Dynamic metasurface for broadband electromagnetic modulator in reflection. Paper presented at the 2016 IEEE 29th international conference on micro electro mechanical systems (MEMS)
15. Yu N, Genevet P, Kats Mikhail A, Aieta F, Tetienne J-P, Capasso F, Gaburro Z (2011) Light propagation with phase discontinuities: generalized laws of reflection and refraction. Science 334(6054):333–337. https://doi.org/10.1126/science.1210713

Chapter 9
Reconfigurable Metasurfaces for Dynamic Polarization Control

9.1 Introduction

Polarization is a property of electromagnetic waves that specifies the orientations of the electrical and magnetic fields, respecting the propagation directions. Many devices, e.g., waveplates, have been proposed to convert the electromagnetic waves' polarization states into desired ones for vast applications, including beamforming [1, 2], optical tweezers [3, 4], optical isolation [5, 6], telecommunications [7, 8], etc. Metasurfaces are arrays of subwavelength meta-molecules designed to control the incident electromagnetic waves' phase, amplitude, and polarization states. The incident wavefront can be converted to arbitrary forms by the predesigned spatial arrangement of meta-molecules. Metasurfaces have extraordinary electromagnetic properties that cannot be realized in any natural materials due to scattering or coupling between the meta-molecules [9–11]. As a result, the electromagnetic properties of the metasurfaces are highly dependent on the geometrical symmetries of the meta-molecules, most of which can only work under specific polarized incidence, e.g., linear-polarized waves or circular-polarized waves [12–14]. In most meta-molecules' designs, the phase profiles of orthogonal-polarized incidences, e.g., left and right circular polarized waves, are different, resulting in 50% diffraction efficiency lost for unpolarized incidence. For example, meta-molecules based on Pancharatnam-Berry (PB) phase shifters the opposite phase changes with left and right circular polarized incidences. Therefore, it is demanding to accommodate each polarization state individually, which inevitably limits the versatilities and tunability of the metasurface-based devices.

On the other hand, microfluidic metasurfaces have more flexibility in the phase profile controls, which are enabled by the symmetry tuning of the meta-molecules. The individually tuning of the meta-molecules can be applied to the metasurfaces with reconfigurable output wavefronts, enabling meta-devices with switchable functionalities. More importantly, incidences with different polarization states can be

W. Zhu and A.-Q. Liu, *Metasurfaces: Towards Tunable and Reconfigurable Meta-devices*, Microfluidics and Nanophotonics: Science and Engineering 1,
https://doi.org/10.1007/978-981-19-6925-6_9

selectively controlled by rationally designed meta-molecules composed of liquid-state materials. This chapter discusses microfluidic metasurfaces for arbitrary and independent control of incident electromagnetic waves with orthogonal polarization states.

9.2 Microfluidic Metasurfaces with Reconfigurable Cross-Shaped Meta-molecules

The polarization states of the electromagnetic waves offer other degrees of freedom for wireless/fiber communications [15], holographic imaging [16], quantum computing [17], etc. Recently, metasurfaces enabled flat electromagnetic devices are widely demonstrated for polarization states conversion [18], polarization-dependent beam splitting [19] and holographic imaging [20, 21], optical chirality [22, 23], etc. However, most anisotropic meta-molecules have different coupled electromagnetic responses to incident waves with orthogonal polarization states, which greatly limits the design flexibilities and applications of the polarization-dependent devices based on metasurfaces. This section discusses a cross-shaped meta-molecule based on electric resonance, which has independent responses to the incident waves with two orthogonal polarization states.

The cross-shaped meta-molecules are shown in the inserts of Fig. 9.1, which have two orthogonal symmetry axes, i.e., x- and y-axis. The cross-shaped meta-molecule comprises three layers, which, from top to bottom, are the PDMS layer, the quartz spacing layer, and the copper ground layer. Therefore, the cross-shaped meta-molecules are designed to work in the reflection mode with the transmitted electromagnetic waves eliminated by the copper ground layer. The PDMS layer is to form and control the liquid–metal cross using a microfluidic control system. The liquid–metal cross consists of two rectangular arms with longer sides along x- and y-directions denoted as x- and y-arm, respectively. The liquid metal is injected into two identical rectangular microchannels perpendicular to each other. Here, the widths of the microchannels W are much smaller than their lengths, resulting in the decoupling of the electrical modes excited by the incident waves polarized along the x- and y-directions. In the incident frequency region, the electrical resonances can only be excited along the longer side of the liquid–metal arm, which has a trivial response to the incident electrical fields along the orthogonal direction. In the meantime, the lengths of x- and y-arm can be individually tuned by changing the air pressures applied to the four air inlets, which results in the tuning of the electrical resonance modes and the phase shifts of the incidence with the electrical fields along the arms. As a result, the phase shifts of x- and y-polarized incidence are independently controlled by the arm lengths along x and y-directions, i.e., L_x and L_y, respectively. Here, the electrical fields of the incident waves propagate along the z-direction can be written as the following by ignoring the initial phase difference between x- and y-polarized incidence.

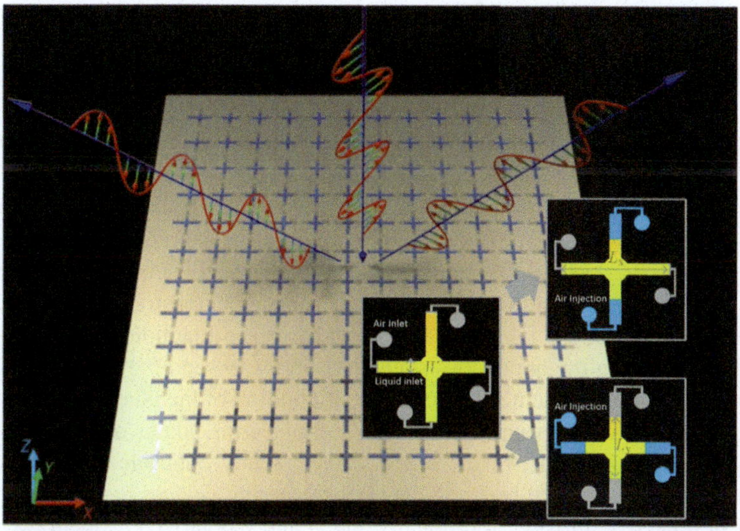

Fig. 9.1 The schematic of a metasurface based on cross-shaped meta-molecules. The linearly polarized incidences excite the electrical resonances within the cross-shaped meta-molecules along their two orthogonal arms along the x- and y-directions. The inserts show the tuning mechanisms and the zoomed-in view of the meta-molecule with a period of 7.5 mm and the arm length up to 7 mm. The yellow color represents the liquid metal

$$\begin{cases} E_x(-z) = A_x e^{i(-kz-\omega t)} \\ E_y(-z) = A_y e^{i(-kz-\omega t)} \end{cases} \tag{9.1}$$

where ω is the angular frequency of the incidence, k is the wave vector of the incident waves, and A_x and A_y are the amplitudes of the incident x- and y-polarized electrical fields, respectively.

The reflected x- and y-polarized electrical fields E_{rx} and E_{ry} can be expressed as the following when the metasurface locates at the $z = 0$ plane.

$$\begin{cases} E_{rx}(z) = E_x(0)e^{i(kz-\omega t+\phi_x)} \\ E_{ry}(z) = E_y(0)e^{i(kz-\omega t+\phi_y)} \end{cases} \tag{9.2}$$

where ϕ_x and ϕ_y are the phase shifts induced by x- and y-arms, respectively.

Based on the discussion of the generalized law of refraction and reflection in Chap. 3, the phase gradients determine the reflection angles of the reflected x- and y-polarized EM waves under normal incidence, which can be described as

$$\begin{cases} \theta_x = \sin^{-1}\left(\frac{\lambda}{2\pi} \cdot \frac{d\phi_x}{dr}\right) \\ \theta_y = \sin^{-1}\left(\frac{\lambda}{2\pi} \cdot \frac{d\phi_y}{dr}\right) \end{cases} \tag{9.3}$$

where θ_x and θ_y are the reflection angles of x- and y-polarized incidence, respectively; λ is the wavelength of the incident wave, and $r = \sqrt{x^2 + y^2}$.

The reflection angles of x and y-polarized incidence can be individually controlled when ϕ_x and ϕ_y are not coupled to each other. In other words, the meta-molecules respond to the two orthogonal polarized incident waves independently. More importantly, multiple functionalities of the metasurface are demonstrated based on cross-shaped meta-molecules. For example, the metasurface functions as a polarization beam splitter when the anomalous reflection angles (θ_x and θ_y) are different. The microfluidic control system's dynamic reconfiguration of the phase gradient can be applied to the beam steering devices. The beam steering angles of the two orthogonally polarized incident waves can be independently controlled, which is quite challenging to be realized by traditional beam steering devices based on refractive lenses or reflective mirrors. Additionally, the x- and y-polarized incident waves can be reflected in the same direction when the phase gradients of x and y-polarized incidence are identical, i.e., $\frac{d\phi_x}{dx} = \frac{d\phi_y}{dx}$ and $\frac{d\phi_x}{dy} = \frac{d\phi_y}{dy}$. In the meantime, the phase shift (ϕ_x and ϕ_y) can be different, e.g. $\phi_x = \phi_y \pm 90°$. As a result, Eq. 9.2 can be rewritten as the following when $E_x(0) = E_y(0) = E_0$,

$$\begin{cases} E_{rx}(z) = E_0 e^{i(kz-\omega t)} \\ E_{ry}(z) = E_0 e^{i(kz-\omega t \pm 90^\circ)} \end{cases} \tag{9.4}$$

In this case, the reconfigurable metasurface functions as a quarter waveplate, which converts a linearly polarized incidence to a circularly polarized one. As a result, the metasurfaces have multiple functionalities, which can be dynamically switched from one to the another.

The cross-shaped meta-molecule consists of two rectangular liquid metal slabs, a SiO_2 spacing layer, and a copper ground layer. The phases of reflected electromagnetic waves are shifted by the electrical resonances of the liquid metal slabs, which function as the slab antennas. The length of the liquid metal slab determines the induced resonance modes of electron currents with the resonant wavelengths proportional to the arm lengths, which is similar to a dipole antenna. No resonance modes can be excited when the arm length is much smaller than the wavelength. Therefore, the width of the liquid metal slabs is chosen to be 0.3 mm so that the electrical resonances can only be excited along the longer side of the liquid metal slab. The electrical field distributions of the meta-molecules are shown in Fig. 9.2a, b, when the arm lengths along the x-direction L_x are fixed at 7 mm and the arm lengths along the y-direction L_y are 2 mm and 7 mm, respectively. The electrical field of the incident wave is along the x-direction with an incident frequency of 15 GHz. The liquid metal slabs oriented along the x- and y-directions are denoted as x- and y-arm, respectively. The dipole resonances are excited on x-arms of both meta-molecules, as shown in Fig. 9.2a, b, which are identical when the y-arm is tuned from 2 to 7 mm. No resonance is observed on the y-arms. It can be concluded that the incident wave excites the electrical resonances only on the arms oriented along the direction of its electrical field, and the length of one arm of the liquid metal cross has a trivial effect on the electrical resonances excited on the other. In other words, the electrical resonances induced by the orthogonally polarized incidences are not coupled, which can be tuned independently.

The phase and amplitude contour maps as functions of L_x and L_y also prove the individually tuning of the two orthogonally-polarized incidences by the cross-shaped meta-molecules. The incident electrical field is along the x-direction, and the color represents the phase and amplitude of the reflection, as shown in Fig. 9.2c, d, respectively. Both the phase and amplitude of the reflection are independent of L_y, indicating that the tuning of the arm along the y-direction has a trivial effect on the reflected waves with electrical fields along the x-direction. A similar conclusion

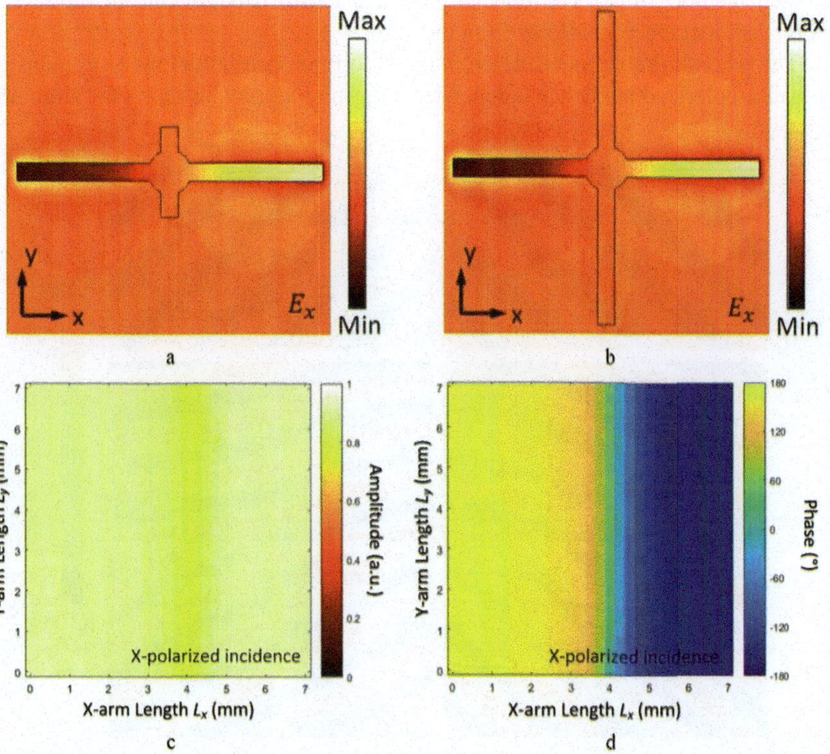

Fig. 9.2 Simulation results of the cross-shaped meta-molecules when the incident frequency is 15 GHz. **a** and **b** Show the electrical field distributions of the meta-molecules when the incident wave is polarized along the x-direction. The arm lengths L_x are fixed at 7 mm, while L_y is tuned from (a) 2 mm to (b) 7 mm. **c** and **d** Show the contour maps of reflected (c) amplitude and (d) phase as functions of L_x and L_y

can be made for the incidence with electrical field along the y-direction due to the rotational symmetry of the metasurface design. Figure 9.2c shows an area of low reflectivity when L_x ranges from 3.4 to 4.5 mm. This area is carefully avoided during the design and tuning processes. The reflectivity is above 90% elsewhere. As shown in Fig. 9.2d, the phase distribution shows 2-π phase shift can be achieved by changing the arm lengths of the cross-shaped meta-molecules.

Figure 9.3 shows the contour maps of the electrical fields reflected from the reconfigurable metasurface with three different configurations. With a frequency of 15 GHz, the incident electromagnetic waves propagate along the z-direction, which is perpendicular to the metasurface. The left and right columns show the simulation results of y- and x-polarized incidences, respectively. The electrical fields are represented by the color, while the wavefronts of the reflections can derive the reflection angles. Here, the propagating directions of the reflected electromagnetic waves are highlighted by the red arrows, and the black-dashed lines represent the normal of the metasurface plane. The simulations are conducted by using the finite element method (FEM). The reflected electrical field distributions are the subtractions of the incident electrical fields from the total.

The reconfigurable metasurface is tuned to three configurations to demonstrate x- and y-polarized incidence independent control. Figure 9.3a shows the simulation

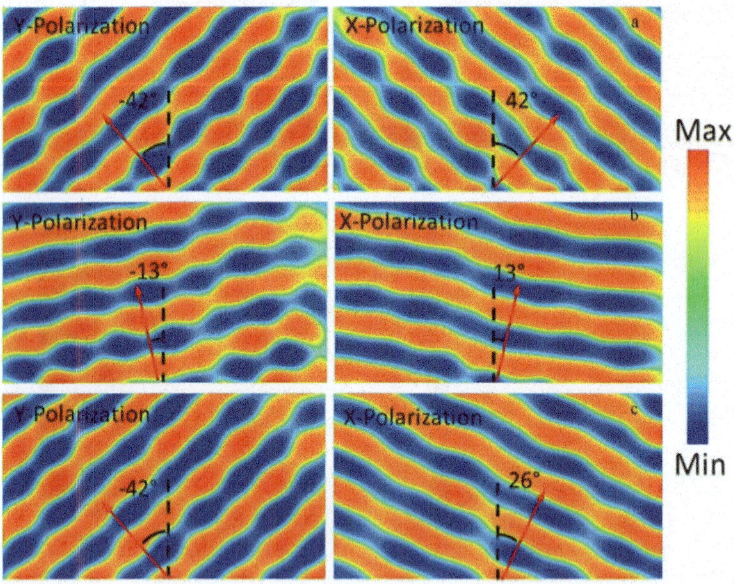

Fig. 9.3 Simulation results of the electrical fields reflected from the reconfigurable metasurface with normal incidence. **a** The metasurface is tuned to have an opposite phase gradient for x- and y-polarized incidences, i.e., $d\Phi_x/dx = -d\Phi_y/dx$. The lengths of the supercells for x- and y-polarized incidences are both 30 mm. **b** The lengths of the supercell are both changed to 90 mm. **c** The lengths of the supercells for x- and y-polarized incidences are 45 mm, and 30 mm, respectively

results of the first configuration with $\frac{d\phi_x}{dx} = -\frac{d\phi_y}{dx}$, and the supercell length $\xi =$ 30 mm. The reflected waves are diffracted to the -1st and 1st order of the 30-mm-period metasurface grating for y- and x-polarized incidence, respectively, while the anomalous reflection angles are -42° and 42°, respectively. The reconfigurable metasurface functions as a polarization beam splitter at this configuration with a splitting angle of 84°. For the second configuration, the supercell lengths ξ of the metasurface are tuned to 90 mm for both x- and y-polarized incidence. The anomalous reflection angles for x- and y-polarized incidence are changed to 13° and − 13°, respectively. As a result, the splitting angle is tuned from 84° to 26° by changing the metasurface configuration. The different phase gradient of x- and y-polarized incidences is achieved when the reconfigurable metasurface is tuned to the third configuration, as shown in Fig. 9.3c. The supercell lengths for x- and y-polarized incidences are tuned to 45 and 30 mm, respectively, resulting in the different phase gradients, i.e., $\frac{d\phi_x}{dx} \neq -\frac{d\phi_y}{dx}$. The anomalous reflection angles are tuned to 26° and − 42° for x- and y-polarized incidence, proving the independent tuning of different polarization states.

Figure 9.3 only shows the simulation results of the reconfigurable metasurface with phase gradient along the x-direction. Theoretically, the reflected waves can be directed to any omnidirectional, with the arbitrary reconfiguration of the metasurface's phase gradient. However, the choices of the beam steering angles are limited by the finite sizes of the meta-molecules, i.e., the supercell lengths have to be integer multiple of the metasurface period (7.5 mm in this case). Additionally, there are some differences observed in the electrical field distributions for x- and y-polarized reflections, which is due to the grating effect of the metasurface.

The reconfigurable metasurface is fabricated by using the soft lithography processes discussed in Chap. 3. First, the photoresist SU-8 100 is spin-coated on a silicon wafer. The thicknesses of the silicon wafer and the SU-8 photoresist are 0.6 mm and 0.1 mm, respectively. Then, UV light exposure is applied to define the pattern of the microfluidic channels. After the post-baken, the exposed sample is placed into the SU-8 developer for development. The PDMS was poured on the SU-8 mold and peeled off after it became solid. In this work, two different PDMS layers are fabricated. One is with cross-shaped channels for the liquid metal cross-shaped meta-molecules. The other is designed for the liquid and air channels with the control system. The two PDMS layers are bonded together with plasma treatment. Finally, a copper layer is sputtered to a glass wafer as the PEC layer and bonded with the PDMS layer.

The fabricated reconfigurable metasurface consists of three layers, as shown in Fig. 9.4. The top layer is the PDMS layer with the control system and the cross-shaped reservoirs for the liquid metal meta-molecules. The arm's length, width, and height are 7 mm, 0.4 mm, and 0.1 mm, respectively. Therefore, the liquid metal meta-molecules can have their arm length dynamically tuned from 0.4 mm to 7 mm. There are two different microchannels for air control and liquid metal delivery, which have widths of 0.1 mm and 0.25 mm, respectively. The pumping pressure of the liquid metal is carefully controlled to prevent them from entering the air channels. In the

Fig. 9.4 Photographs of the fabricated reconfigurable metasurface. The inserts show Zoom-in views of a meta-molecule with the x-arm and the y-arm tuned with different lengths by injecting air through the four air inlets. The liquid metal is highlighted with false yellow color

meantime, the air can be pumped into the cross-shaped reservoirs due to its low viscosity. The overall height of the PDMS layer is 2 mm, which is composed of a 1-mm microfluidic control system and a 1-mm liquid metal reservoir array with a period of 7.5 mm. The 1-mm Borofloat glass wafer is the spacing layer between the copper ground plate and the PDMS layer. Here, the copper layer can be either sputtered onto the glass layer or realized by mounting a copper plate with the glass wafer. In this photograph, the copper plate is removed from the glass wafer so that the liquid metal cross-shaped meta-molecules can be clearly observed. The inserts show the independent tuning for x- and y-arms of a meta-molecule with the liquid metal highlighted with the false yellow color.

The experimental setup for far-field directivity measurement of the reflective reconfigurable metasurface is shown in Fig. 9.5a. Here, two wideband double-ridged horn antennas (HD-20180DRHA10SK) are connected to a vector network analyzer (Agilent N5230A), which functions as the source (green) and the receiver (red). The source antenna is fixed at the normal of the metasurface plane, and the incident electromagnetic waves propagate along the z-direction. The distance between the source antenna and the metasurface is set to be 1.2 m so that the metasurface is placed on the waist of the incident Gaussian beam to mimic the plane wave incidence. The receiving antenna is mounted on a circular track with a radius of 1.5 m to measure the directivities with the reflection angle ranging from $-90°$ to $90°$.

In experiment, the metasurface is tuned to three different configurations, as discussed in Fig. 9.3. The experimental results of dynamic polarization beam splitting are shown in Fig. 9.5b, c. The anomalous reflection angles of x- and y-polarized incidences are 42° and $-42°$, respectively when the supercell lengths $\xi = 30$ mm

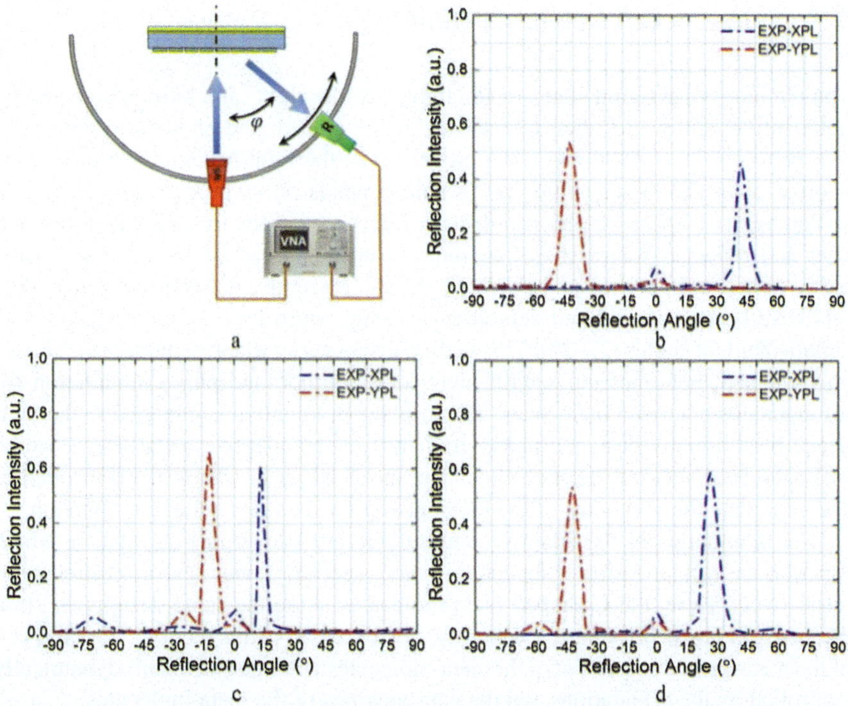

Fig. 9.5 Experimental setup and results of the reconfigurable metasurface **a** the experimental setup for the far-field measurement of a reflective metasurface. The experimental results of tunable beam splitting when the splitting angle is tuned from 84° (**b**) to 26° (**c**). **d** shows the independent beam steering for the x- and the y-polarized incidence with the steering angles of 26° and −42°, respectively. The red and blue lines represent the y- and x-polarized incidence, respectively

and $\frac{d\phi_x}{dx} = -\frac{d\phi_y}{dx}$. The measured diffraction efficiencies of x- and y-polarized incidences are 45.9% and 53.1%, respectively. The beam splitting angle is tuned from 84° to 26° when the supercell length is changed to 90 mm, as shown in Fig. 9.5c. The measured diffraction efficiencies of x- and y-polarized incidences are changed to 59.6% and 64.3%, respectively. The independent beam steering for x- and y-polarized incidences are shown in Fig. 9.5d, when the supercell length for x- and y-arms are tuned to 45 mm and 30 mm, respectively. The anomalous reflection angles of x- and y-polarized incidences are measured to be 26° and − 42°, respectively, and the diffraction efficiencies are 59.0% and 53.6%, respectively.

The experimental results agree well with the simulation results, as shown in Fig. 9.3, which proves the independent control of the anomalous reflection angles for the orthogonally-polarized incidences using the reconfigurable metasurface. The diffraction efficiencies decrease when the reflection angle increases, which can be explained by the shadow effect of the meta-molecules.

9.3 Spin-Locked Retroreflection

The circular polarization states of the light, i.e., the right- and left-handed circular polarization states, are often referred to as the spin of the photons to create the link between the particle and wave properties of the light beam [24]. Same as the circular polarization states, the light's spin depends on its propagation directions, switching after reflection. Retroreflection refers to the phenomena when the incident and reflected waves share the same beam path, widely demonstrated using bulky optics, e.g., prisms and mirrors [25, 26]. Recently, flat and thin retroreflectors have been proposed and demonstrated using multi-layered metasurfaces with certain incident angles [27–29]. This section discusses an adaptive metasurface based on C-shaped meta-molecules with identical amplitude and phase modulation for orthogonally-polarized incidences. The in-plane momentum is imparted to the incidence by rationally designed phase gradience of the adaptive metasurface, which locks the spin state of the incidence. The metasurface is adaptively controlled to accommodate the variations of the incident angle with a retroreflection function.

The schematics of the adaptive metasurface are shown in Fig. 9.6a, b, which function as a polarization-independent beam steering device and a retroreflector, respectively. Here, the C-shaped meta-molecules share the same designs as those discussed in Fig. 8.1 of Chap. 8, which are not discussed in this section. The supercell of the metasurface consists of eight meta-molecules with the phase shift dynamically controlled by the orientations and the gap-openings of the meta-molecules.

Polarization states of the electromagnetic waves in free space can be described by using Jones vectors. The electrical fields of the light propagating along the z-direction can be written as the following.

$$\begin{pmatrix} E_x(t) \\ E_y(t) \end{pmatrix} = \begin{pmatrix} E_{0x} \\ E_{0y} \end{pmatrix} e^{i(kz-\omega t)} \tag{9.5}$$

where k and ω are the wave vector and the angular frequency of the incidence, respectively. The Jones vector J is defined as the following.

$$J = \begin{pmatrix} E_{0x} \\ E_{0y} \end{pmatrix} \tag{9.6}$$

The Jones matrix is an operator acting on the Jones vectors to describe the output electromagnetic waves from devices such as beam deflectors, polarization beam splitters, lenses, mirrors, etc. Therefore, the reflected waves from a metasurface can be written as the following.

$$\begin{pmatrix} E_x \\ E_y \end{pmatrix} = \begin{pmatrix} R_{xx} & R_{xy} \\ R_{yx} & R_{yy} \end{pmatrix} \begin{pmatrix} E_{0x} \\ E_{0y} \end{pmatrix} \tag{9.7}$$

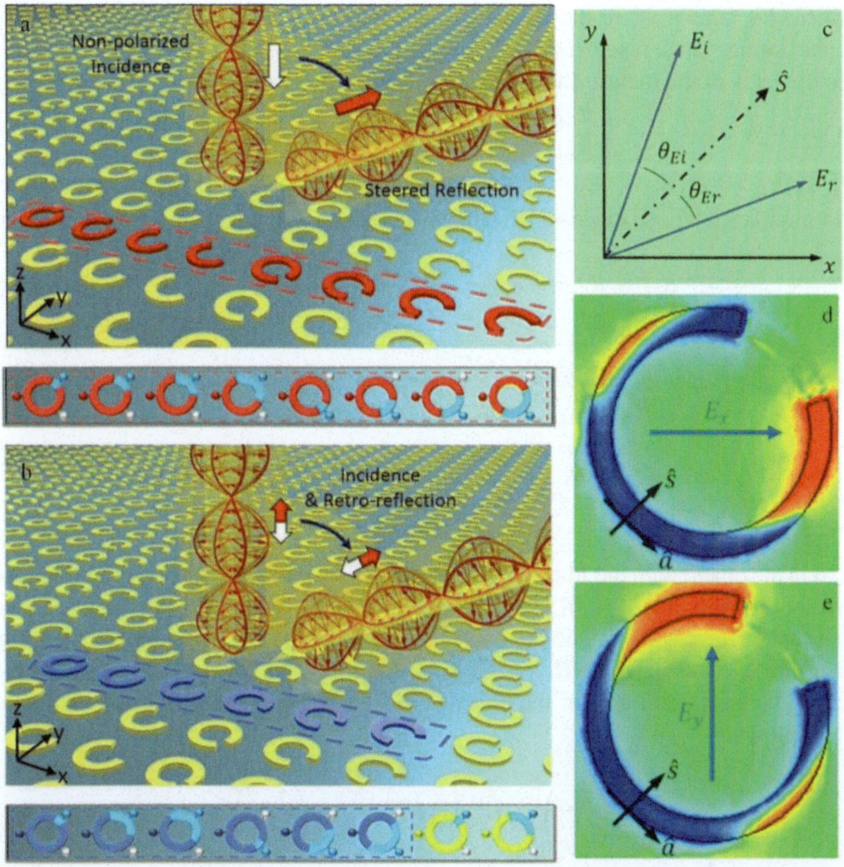

Fig. 9.6 Schematics and working principle of adaptive metasurface for spin-locked retroreflection. **a** Polarization-independent beam steering and **b** spin-locked retroreflection. The inserts show super-cell tuned with eight meta-molecules. **c** Linearly-polarized incidence and reflection with the angle θ_{Ei} and θ_{Er}, respectively, between the electrical field orientation direction and the meta-molecule's symmetric axis *S*. The contour maps of electrical field distributions of meta-molecules oriented along 45° when the incident electric field is along with the **d** x-direction and **e** y-direction

where $R_{LP} = \begin{pmatrix} R_{xx} & R_{xy} \\ R_{yx} & R_{yy} \end{pmatrix}$ is the Jones matrix for linearly polarized electromagnetic waves. R_{xx}, R_{xy}, R_{yx}, and R_{yy} are the complex reflection coefficients of the metasurface.

Similarly, the electrical fields of left- and right-polarized reflected waves EL and ER can be written as follows for circularly polarized incidence.

$$\begin{pmatrix} E_L \\ E_R \end{pmatrix} = \begin{pmatrix} R_{LL} & R_{LR} \\ R_{RL} & R_{RR} \end{pmatrix} \begin{pmatrix} E_{0L} \\ E_{0R} \end{pmatrix} \tag{9.8}$$

where E_{0L} and E_{0R} are the electrical fields of the incidence with left- and right-polarization states, respectively; R_{LL}, R_{LR}, R_{RL}, and R_{RR} are the complex reflection coefficients of the metasurface. The Jones matrix for circular polarized electromagnetic waves is $R_{CP} = \begin{pmatrix} R_{LL} & R_{LR} \\ R_{RL} & R_{RR} \end{pmatrix}$. And E_L and E_R can be decomposed to E_x and E_y as $E_L = \frac{1}{\sqrt{2}}(E_x + iE_y)$ and $E_R = \frac{1}{\sqrt{2}}(iE_x + E_y)$. Therefore, the Jones matrix for circular polarized electromagnetic waves can be written as the following.

$$R_{CP} = \frac{1}{2}\begin{pmatrix} R_{xx} + R_{yy} + iR_{yx} - iR_{xy} & -iR_{xx} + iR_{yy} + R_{yx} + R_{xy} \\ iR_{xx} - iR_{yy} + R_{yx} + R_{xy} & R_{xx} + R_{yy} - iR_{yx} + iR_{xy} \end{pmatrix} \tag{9.9}$$

For a meta-molecule with fourfold rotational symmetry or polarization independence, i.e., $R_{xx} = R_{yy}$ and $R_{xy} = R_{yx}$, the R_{cp} can be written as,

$$R_{CP} = \begin{pmatrix} R_{xx} & R_{xy} \\ R_{yx} & R_{yy} \end{pmatrix} = R_{LP} \tag{9.10}$$

On the other hand, the Jones matrix becomes diagonal and anti-diagonal when the meta-molecules have 0% ($R_{xy} = R_{yx} = 0$) and 100% ($R_{xx} = R_{yy} = 0$) polarization conversion, respectively. The C-shaped meta-molecules have near-unity polarization conversion efficiency, i.e., $R_{xy} = R_{yx} = 1$ and $R_{xx} = R_{yy} = 0$, ignoring the ohmic loss. The electrical fields of the reflected electromagnetic waves can be written as the following.

$$\begin{pmatrix} E_L \\ E_R \end{pmatrix} = \begin{pmatrix} 0 & 1 \\ 1 & 0 \end{pmatrix}\begin{pmatrix} E_{0L} \\ E_{0R} \end{pmatrix} = \begin{pmatrix} E_{0R} \\ E_{0L} \end{pmatrix} \tag{9.11}$$

For retroreflection, the propagation direction of the incidence is opposite to that of the reflected waves, resulting in the switching between the two orthogonal polarization states. Therefore, Eq. 9.11 can be rewritten as,

$$\begin{pmatrix} E_L \\ E_R \end{pmatrix} = A\begin{pmatrix} E_{0L} \\ E_{0R} \end{pmatrix} \tag{9.12}$$

where A is the attenuation coefficient due to the ohmic loss. Furthermore, the retroreflection of the adaptive metasurface reserves the spin/circular-polarization states of the incidences.

The cross-polarization reflection of the C-shaped meta-molecules is shown in Fig. 9.6c when the symmetry axis of the meta-molecule is at a 45-degree angle with the x-axis. The incident electrical field E_i and the reflected electrical field E_r are located at both sides of the symmetry axis $\boldsymbol{S}$ with the same separation angle, indicating a cross-polarization reflection. The electrical field distributions for x- and y-polarized incidences are shown in Fig. 9.6d, e, respectively. The identical electrical field distributions for orthogonally-polarized incidences prove that the C-shaped meta-molecules have polarization-independent electromagnetic responses, which is a perfect candidate for the spin-locked metasurface. Detailed simulation results of the C-shaped meta-molecules can be found in Chap. 8, which are not discussed in this section.

The adaptive metasurface is fabricated by using soft-lithography processes, as shown in Fig. 9.7. Here, the C-shaped meta-molecules with a radius of 4 mm are arranged in a square lattice array with a period of 5 mm. The height of the microfluidic channel is 0.1 mm for both liquid metal and air channels. The PDMS layer is 2-mm thick which is bonded on a 1.25-mm thick PDMS plate. A copper ground plate is mounted on the backside of the PDMS substrate, which is not shown in the photograph. In the experiment, the Galinstan is used as the liquid metal. The hydrochloric vapor is pumped into the air channels to regulate the liquid–metal meta-molecules and clean the microfluidic channels. The orientation of the meta-molecules can be switched between 45° and 135°, and the gap opening of the meta-molecules can be tuned from 5° to 180°.

The experimental setup of the metasurface retroreflector, as shown in the insert of Fig. 9.7, is similar to that of the reflective reconfigurable metasurface, as shown in Fig. 9.5a. Here, both the source and receiver antennas are mounted on the 1.2-m circular track to measure the retroreflection. The source antenna only generates linearly-polarized incidences, which is rotated 90° to generate the incidences with x- and y-polarization states. In the meantime, the directivities of both co- and cross-polarized reflections are measured by rotating the receiver antenna. In the experiment, the reflection of 15-GHz incidences is measured at different incident angles, i.e.,

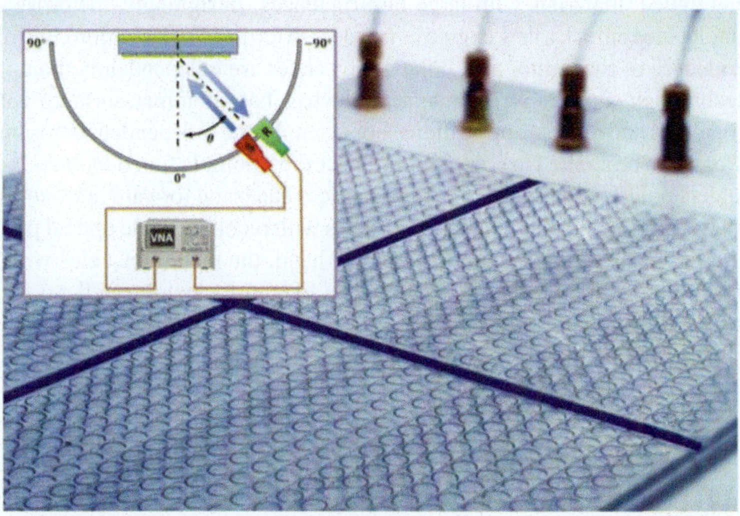

Fig. 9.7 An overview photograph of the fabricated adaptive metasurface. The insert shows the experimental setup for the spin-locked retroreflector

10°, 12°, 15°, and 20°, with different metasurface configurations. The measured experimental results of linearly-polarized incidences and reflections are converted to circular polarization by using Eq. 9.9.

The experimental results of the retroreflector are shown in Fig. 9.8. The x- and y-polarization states are denoted as TE- and TM-polarization states, respectively. The reflection intensities and efficiencies are normalized with the reflection measured under the normal incidence, while all the meta-molecules are tuned to be identical for the retroreflection. Figure 9.8a shows the reflection intensities as functions of the reflection angle with different incident angles, i.e., 10°, 12°, 15°, and 20°. The adaptive metasurface is tuned to different configurations with different supercell lengths to accommodate the change of the incident angle, as shown in Fig. 9.8b. The measured reflection intensity decreases as the retroreflection angle increases, which is also verified by the measured reflection efficiencies, as shown in Fig. 9.8c. The deceasing of the reflection efficiencies is mainly due to the shadow effect of the meta-molecules. The size of the meta-molecules is approximately a quarter of the wavelength, which is dominated by the Mie scattering. The measured phase difference between TE- and TM-polarized reflection is less than 0.033π, indicating a trivial change in the circular polarization states of the incidences, which proves the retroreflection by the adaptive metasurface is spin-locked.

9.4 Summary

The symmetry breaking of the meta-molecules has now been intensively applied to artificial optical materials with large birefringences. Intriguingly, the cross-shaped meta-molecules unlock a new degree of freedom in terms of polarization control when the incidences of two orthogonal polarization states are independently manipulated. This technology supplies electromagnetic devices based on metasurfaces with new functionalities and diverse applications. For example, the independent tailoring of the wavefronts can be applied to holographic devices for multiple functionalities based on the incident polarization states. The metasurfaces can bring forward new methods for information processing based on metasurfaces with reconfigurable optical properties for different polarization states. On the other hand, the meta-molecules with cross-polarization reflection have now been applied to spin-locked retroreflection, which can replace the bulky retroreflectors based on refractive and reflective optics also of great interest to quantum optics devices.

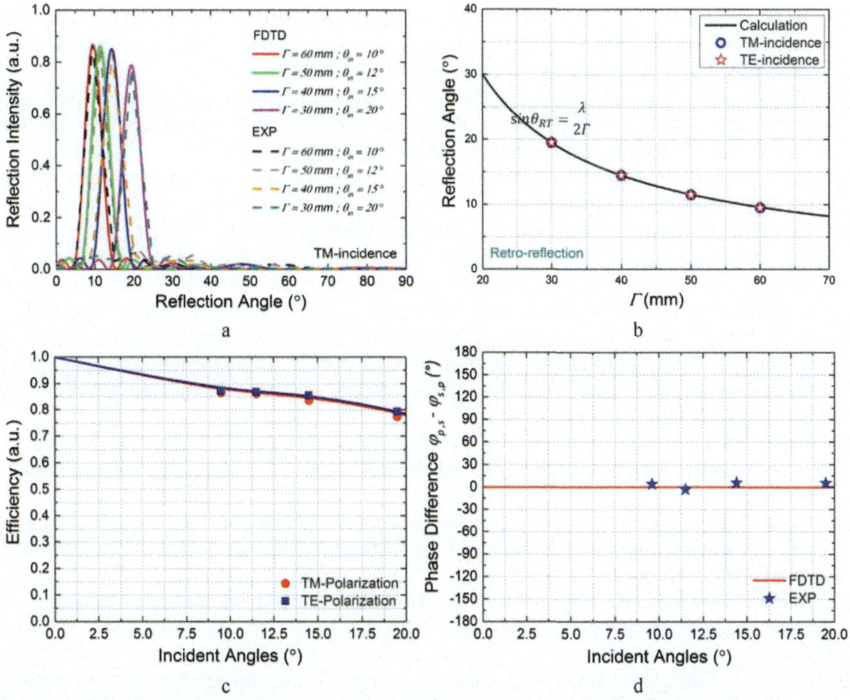

Fig. 9.8 Experimental results of the retroreflector based on the adaptive metasurface. **a** Cross-polarized reflection intensities with different incident angles for TM-polarized (y-polarized) incidences are normalized by the normal incidence reflection. **b** Relation between the angles of retroreflection and the supercell length of the metasurface. **c** The reflection efficiency at various angles for TM-(red-sphere) and TE-polarized (blue-cube) incidences. **d** The phase difference between retroreflections under TE- (s-) and TM- (p-) polarized incidences at various angles. Reproduced with permission [29] copyright from 2018 Wiley–VCH

References

1. Veen BDV, Buckley KM (1988) Beamforming: a versatile approach to spatial filtering. IEEE ASSP Mag 5(2):4–24. https://doi.org/10.1109/53.665
2. Chen JC, Kung Y, Hudson RE (2002) Source localization and beamforming. IEEE Signal Process Mag 19(2):30–39. https://doi.org/10.1109/79.985676
3. Moffitt JR, Chemla YR, Smith SB, Bustamante C (2008) Recent advances in optical tweezers. Annu Rev Biochem 77(1):205–228. https://doi.org/10.1146/annurev.biochem.77.043007.090225
4. Guo H, Li Z (2013) Optical tweezers technique and its applications. Sci China Physics, Mech Astron 56(12):2351–2360. https://doi.org/10.1007/s11433-013-5355-3
5. Yu Z, Fan S (2009) Complete optical isolation created by indirect interband photonic transitions. Nat Photonics 3(2):91–94. https://doi.org/10.1038/nphoton.2008.273
6. Bi L, Hu J, Jiang P, Kim DH, Dionne GF, Kimerling LC, Ross CA (2011) On-chip optical isolation in monolithically integrated non-reciprocal optical resonators. Nat Photonics 5(12):758–762. https://doi.org/10.1038/nphoton.2011.270

7. Fatome J, Pitois S, Morin P, Millot G (2010) Observation of light-by-light polarization control and stabilization in optical fibre for telecommunication applications. Opt Express 18(15):15311–15317. https://doi.org/10.1364/OE.18.015311
8. Li X, Voss PL, Chen J, Sharping JE, Kumar P (2005) Storage and long-distance distribution of telecommunications-band polarization entanglement generated in an optical fiber. Opt Lett 30(10):1201–1203. https://doi.org/10.1364/OL.30.001201
9. Yu N, Capasso F (2014) Flat optics with designer metasurfaces. Nat Mater 13(2):139–150. https://doi.org/10.1038/nmat3839
10. Kildishev Alexander V, Boltasseva A, Shalaev Vladimir M (2013) Planar photonics with metasurfaces. Science 339(6125):1232009. https://doi.org/10.1126/science.1232009
11. Glybovski SB, Tretyakov SA, Belov PA, Kivshar YS, Simovski CR (2016) Metasurfaces: from microwaves to visible. Phys Rep 634:1–72. https://doi.org/10.1016/j.physrep.2016.04.004
12. Liu T, Meng Y, Ma H, Wang J, Wang X, Zhu R, Qu S (2022) Generating diverse functionalities simultaneously and independently for arbitrary linear polarized illumination enabled by a chiral transmission-reflection-selective bifunctional metasurface. Opt Express 30(5):7124–7136. https://doi.org/10.1364/OE.452395
13. Wu X, Feng Y, Zhang C, Liu HL (2022) Three-dimensional chiral metasurfaces for circular-polarized anomalous beam steering. Opt Lett 47(7):1794–1797. https://doi.org/10.1364/OL.450390
14. Wang H, Li Y, Han Y, Sui S, Wang J, Qu S (2019) A circular-polarized metasurface planar reflector antenna based on Pancharatnam-Berry phase. Appl Phys A 125(4):274. https://doi.org/10.1007/s00339-019-2575-z
15. Andrews MR, Mitra PP, Carvalho R (2001) Tripling the capacity of wireless communications using electromagnetic polarization. Nature 409(6818):316–318. https://doi.org/10.1038/35053015
16. Colomb T, Dahlgren P, Beghuin D, Cuche E, Marquet P, Depeursinge C (2002) Polarization imaging by use of digital holography. Appl Opt 41(1):27–37. https://doi.org/10.1364/AO.41.000027
17. O'Brien Jeremy L (2007) Optical quantum computing. Science 318(5856):1567–1570. https://doi.org/10.1126/science.1142892
18. Shi Z, Zhu Alexander Y, Li Z, Huang Y-W, Chen Wei T, Qiu C-W, Capasso F. Continuous angle-tunable birefringence with freeform metasurfaces for arbitrary polarization conversion. Sci Adv 6(23):eaba3367. https://doi.org/10.1126/sciadv.aba3367
19. Khorasaninejad M, Zhu W, Crozier KB (2015) Efficient polarization beam splitter pixels based on a dielectric metasurface. Optica 2(4):376–382. https://doi.org/10.1364/OPTICA.2.000376
20. Ni X, Kildishev AV, Shalaev VM (2013) Metasurface holograms for visible light. Nat Commun 4(1):2807. https://doi.org/10.1038/ncomms3807
21. Wan W, Gao J, Yang X (2017) Metasurface holograms for holographic imaging. Adv Opt Mater 5(21):1700541. https://doi.org/10.1002/adom.201700541
22. Gorkunov MV, Antonov AA, Kivshar YS (2020) Metasurfaces with maximum chirality empowered by bound states in the continuum. Phys Rev Lett 125(9):093903. https://doi.org/10.1103/PhysRevLett.125.093903
23. Wang Z, Cheng F, Winsor T, Liu Y (2016) Optical chiral metamaterials: a review of the fundamentals, fabrication methods and applications. Nanotechnology 27(41):412001. https://doi.org/10.1088/0957-4484/27/41/412001
24. Xiao S, Wang J, Liu F, Zhang S, Yin X, Li J (2017) Spin-dependent optics with metasurfaces. Nanophotonics 6(1):215–234. https://doi.org/10.1515/nanoph-2016-0121
25. Lee B, Lehmann K, Taylor J, Yalin A (2014) A high-finesse broadband optical cavity using calcium fluoride prism retroreflectors. Opt Express 22(10):11583–11591. https://doi.org/10.1364/OE.22.011583
26. Parker DH (2005) Multidirectional retroreflector assembly with a common virtual reflection point using four-mirror retroreflectors. Precis Eng 29(3):361–366. https://doi.org/10.1016/j.precisioneng.2004.10.001

27. Arbabi A, Arbabi E, Horie Y, Kamali SM, Faraon A (2017) Planar metasurface retroreflector. Nat Photonics 11(7):415–420. https://doi.org/10.1038/nphoton.2017.96
28. Wong AMH, Christian P, Eleftheriades GV (2017, 4–7 Jan) Binary Huygens' metasurface: a simple and efficient retroreflector at near-grazing angles. Paper presented at the 2017 United States national committee of URSI national radio science meeting (USNC-URSI NRSM)
29. Yan L, Zhu W, Karim MF, Cai H, Gu AY, Shen Z, Liu AQ (2018) 0.2 λ0 Thick adaptive retroreflector made of spin-locked metasurface. Adv Mater 30(39):1802721. https://doi.org/10.1002/adma.201802721

Chapter 10
Tunable and Reconfigurable Flat Optics: An Outlook

Flat and thin optical components are key enabling tools for revolutionary technologies and applications, including adaptive cameras, on-chip microscopes, compact Light Detection and Ranging (LIDAR) devices, etc. [1–3]. Before the demonstrations of flat lenses based on metasurfaces, the term "flat optics" was occasionally applied to optical components based on gradient-index (GRIN) materials and diffractive optics, which are relatively flat and thin, e.g., GRIN lens [4], Fresnel lens [5], filter plates [6], optical windows [7], etc. However, the GRIN optical devices are highly dependent on the availability of natural materials, which require substantial thickness due to their phase accumulations based on the propagation of the light. On the other hand, the diffractive optical elements (DOEs) suffer from the limited working bandwidth and complex patterns due to the dispersion effect and high-order diffraction, respectively. Metasurfaces with rationally designed meta-molecules enable the manipulations of the light in terms of amplitude, phase, and polarization with a subwavelength spatial resolution, which has now been widely used in flat optics with flat surfaces and wavelength-level thickness [8–10]. Despite their fascinating features, flat optics based on metasurfaces have become a distributive technology for optoelectronic devices in terms of designs and fabrication processes. Metasurfaces with complementary metal–oxide–semiconductor (CMOS) compatibility will soon be implemented by optoelectronic industries, which are able to design and fabricate the optical components together with the sensors and electronic circuits, empowering diverse technologies based on currently matured semiconductor foundries.

The tunable and reconfigurable metasurfaces supply flat optics with controllable performances and switchable functionalities, which leads to vast applications, e.g., adaptive wavefront control [11], dynamic beam steering [12], controllable polarization conversion [13], tunable absorption, variable focusing, etc. The tunable and reconfigurable flat optics have merits such as compact sizes, flat surfaces, fabrication compatibilities, etc., and have improved performances in terms of their tunabilities. For example, the tuning speeds of the mechanically actuated flat optics based on metasurfaces have now reached the kHz region due to the lightweight of the metasurfaces [14], which are orders of magnitude faster than those based on bulky optical

W. Zhu and A.-Q. Liu, *Metasurfaces: Towards Tunable and Reconfigurable Meta-devices*, Microfluidics and Nanophotonics: Science and Engineering 1,
https://doi.org/10.1007/978-981-19-6925-6_10

components [15]. More importantly, the design flexibilities are greatly improved by the subwavelength meta-molecules with designable optical properties, which unlock extraordinary optical properties of the metasurfaces, such as optical chirality, birefringence, broadband near-unity absorption, etc., leading to promising advances of sciences and technologies.

10.1 Reconfigurable Metasurfaces for Beam Steering

Beam steering devices have been widely demonstrated by using flat optics based on metasurfaces [16, 17]. Most beam steering devices based on metasurfaces are designed following one of the two paradigms widely applied to geometric optics and phased array antennas. The first paradigm is to direct the incident electromagnetic beams in different directions by rotating the optical components, e.g., the mirrors, lenses, gratings, etc. This paradigm is often used by beam-steering devices driven by mechanical forces and has a low tuning speed. The flat optics based on metasurfaces can significantly improve the tuning speeds due to their lightweight. The other paradigm is to steer the electromagnetic beams based on the diffractions of the phased array antennas, which are widely used in microwave antennas and optical spatial light modulators (SLMs). This method has fast tuning speed but requires complex pixel-by-pixel control systems, significantly improving cost and fabrication difficulties.

Figure 10.1a shows a beam steering device based on microfluidic metasurfaces working at the microwave frequency region (Ku band), which is discussed in Chap. 8 [18]. The microfluidic metasurfaces offer great flexibilities of wavefront reconfiguration, which have been demonstrated with switchable functionalities, adaptive dispersion and incident angle control, etc. The applications of such beam steering devices are still quite limited for the following reasons. Firstly, the microfluidic control system has a very slow tuning speed due to the inertia of the micro/nano droplets regulated within the microchannel, which takes seconds, even minutes, to tune a microfluidic metasurface from one configuration to another. Secondly, the beam steering angle cannot be continuously tuned due to the finite sizes of the meta-molecules. Finally, the operating frequencies of the microfluidic metasurfaces are capped at a few THz due to the fabrication limitations of the soft-lithography processes.

Figure 10.1b shows flat optics based on an electrically driven metasurface in the vertical direction with functionalities of variable focusing and beam steering [19]. This beam steering device has the merits of fast tuning speed (> 0.5 kHz), broadband operation, and a simple control system, which unveils the potential fast tuning devices based on mechanical actuation. However, the advantages of the metasurfaces for flat optics are not fully unleashed by this work. The vertical actuation of membrane-like metasurfaces prevents the further improvement of the tuning speed in air ambiance due to the viscosity of the air, which may result in the deformation of the metasurface. Additionally, the wavefront control mechanism has a limited scope of applications, which is difficult to be adapted to the devices with complex output wavefronts.

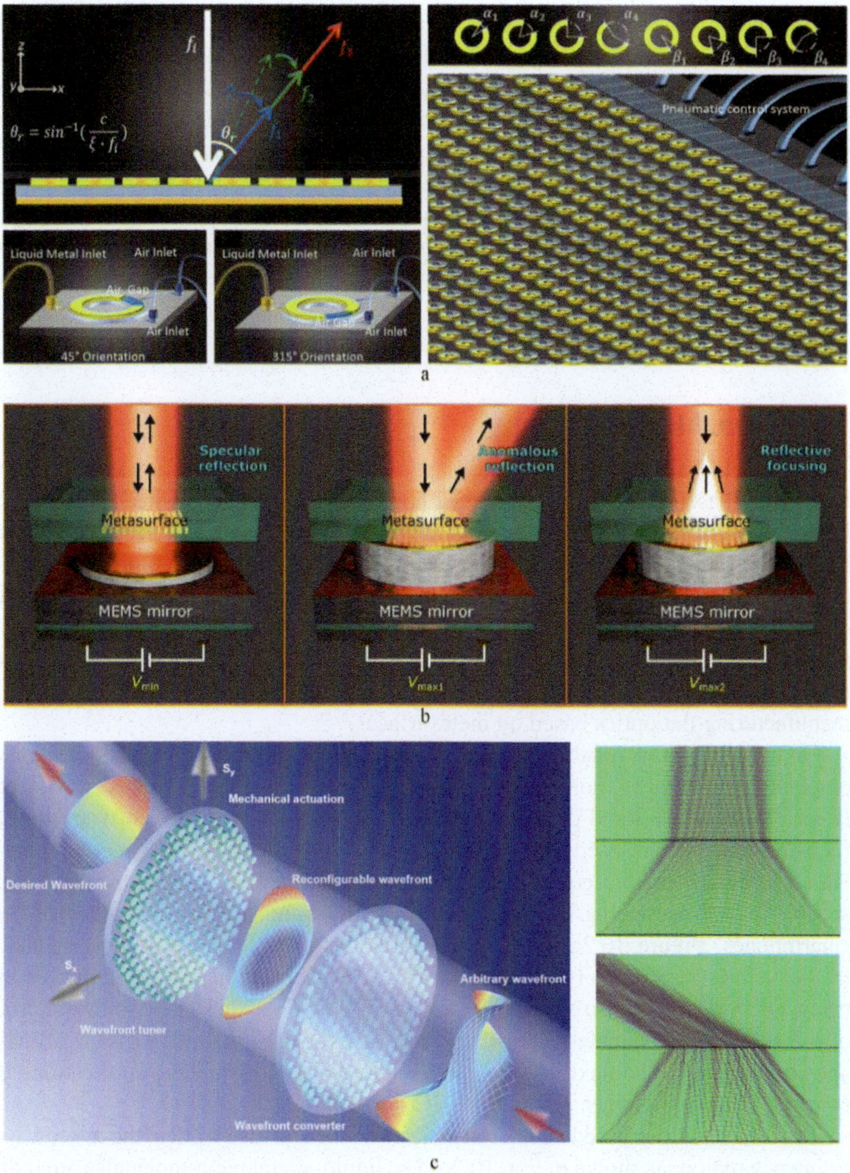

Fig. 10.1 Reconfigurable metasurfaces with beam-steering functionalities. **a** The adaptive metasurface for dispersion controllable beam steering. Reproduced with permission [18] copyright from 2017 AIP Publishing. **b** MEMS metasurface with mechanical actuation in the vertical direction. Reproduced with permission [19] copyright from 2021 American Association for the Advancement of Science. **c** Metasurface wavefront transformer for 6-kHz mechanical beam steering

A metasurface beam steering device based on the in-plane mechanical actuation has recently been demonstrated with a tuning speed of 6 kHz, as shown in Fig. 10.1c [20]. Two piezo actuators move the metasurface along with the two orthogonal directions within the metasurface plane. The metasurface is fabricated on a bulky silicon wafer with a total weight of 2 g, limiting the actuation speed of the piezo actuators. The actuation speed is expected to be 70 kHz by applying the backside thinning process to the metasurface. More importantly, this work came up with a new paradigm for arbitrary wavefront transformation, which supplies the beam steering device with designable tuning speed and can be applied to other wavefront reconfigurable devices, e.g., tunable flat lens.

10.2 Tunable Flat Lens

Flat lenses based on metasurfaces have now been intensively studied due to their intriguing performances and functionalities, e.g., the wavelength-scale thickness, lightweight, flexibility in freeform design, etc. These research efforts aim to commercialize the metasurface flat lenses by improving their working bandwidths, focusing on efficiencies, fabrication processes, etc. For example, the dispersions of the meta-molecules are diligently designed to balance the propagation dispersions for achromatic flat lenses [21]. Now, many new start-up companies focus on designing and manufacturing flat optics based on metasurfaces.

On the other hand, tunable lenses are always bottlenecks for cameras or compact optoelectronic devices. For example, the focal lengths of most camera lenses used in smartphones are fixed due to their compact sizes. As a result, software-based focusing methods are used for the focusing and zoom-in/zoom-out functionalities, which typically require complex artificial intelligence (AI) algorithms for high-speed filming, occupying a large amount of hardware/software resources of the smartphones. Figure 10.2a, b show the variable focal metalenses based on flexible substrates made of polymer materials [22, 23]. This method was first proven by using a PDMS substrate to ease fabrication processes. Then the compositing material of the substrate is changed to acrylate elastomer for the integration of the MEMS actuation system. The flexible-substrate metalens shows promising applications in compact imaging systems.

The tunable lenses based on microfluidic metasurfaces also work in the microwave frequency region, as shown in Fig. 10.2c. The liquid–metal meta-molecules are individually controlled by microfluidic valves, resulting in the reshaping of the output wavefront with dynamic reconfigurations. The metasurface with a complex microfluidic control system randomly accessed to each meta-molecules was first demonstrated in 2015 [24] and named random access reconfigurable metamaterials (RARM). The microfluidic metasurface can be applied to tunable flat lenses and has switchable functionalities due to the capabilities of tuning the individual meta-molecules, resulting in the arbitrary rewriting of the output wavefront. However, the tuning speed is relatively low, which is approximately 1 Hz.

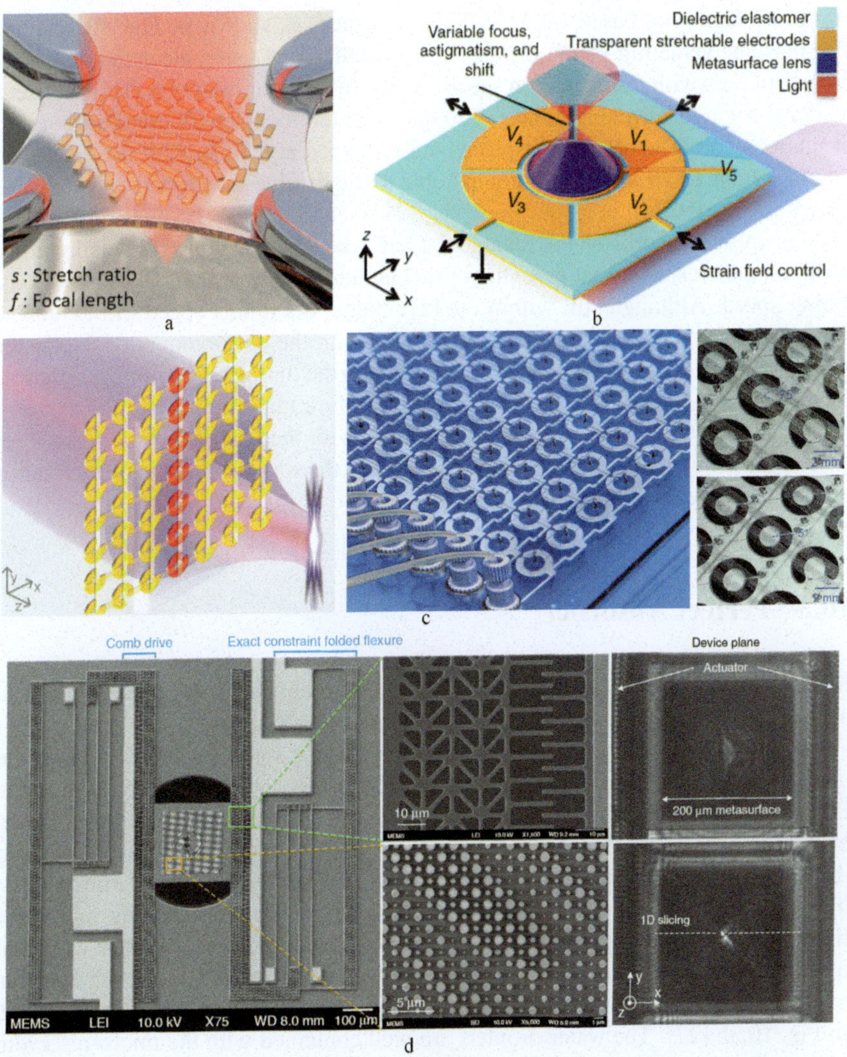

Fig. 10.2 Tunable flat lens based on reconfigurable metasurfaces. **a** and **b** Variable focal lenses based on the stretchable substrates. Reproduced with permission [22] copyright from 2016 American Chemical Society and with permission [23] copyright from 2018 American Association for the Advancement of Science. **c** Tunable flat lens based on a microfluidic metasurface. Reproduced with permission [24] copyright from 2015 Wiley–VCH. **d** Alvarez lens based on a MEMS metasurface. Reproduced with permission [25] copyright from 2020 Springer Nature

An Alvarez lens based on MEMS metasurfaces is shown in Fig. 10.2d [25]. The in-plane actuation of the metasurface is realized by using electrostatic comb-drive actuators with fast actuation speed and low power consumption. This metalens works at a single wavelength, i.e., 1550 nm, with the focal length tuned within tens of micrometers. The uniaxial displacement is capped at 6.3 μm with the maximum tuning voltage of 20 V, which is comparable to the flexible substrate metasurface, as shown in Fig. 10.2b.

Many pioneer works on the tunable flat lens have been demonstrated using metasurfaces, which show great potential in making the on-chip tunable lenses with fast tuning speed. Although still within the kHz range, the tuning speed of the tunable flat lenses is expected to reach MHz or above with the development of the design and fabrication processes, which is comparable to the tuning speed of micro-mirrors driven by state-of-the-art MEMS technologies. More importantly, MEMS metasurfaces' manufacturing processes are compatible with the COMS fabrication widely used in the semiconductor industries, which paves the way for the commercialization of the tunable and reconfigurable flat lenses.

10.3 Perfect Absorber

Broadband and near-unity absorbers are also demonstrated using reconfigurable metasurfaces. A water droplet-based perfect metasurface absorber is shown in Fig. 10.3a, which has a broadband absorption ranging from 8 to 18 GHz [26]. The meta-molecules are water droplets formed by pouring water on different substrates, i.e., paper, glass, polyethylene terephthalate, etc., with surface treatments. This work proves that the water is a good candidate for metamaterial absorbers working in the GHz region where both the real and imaginary parts of water are substantially high. However, the free-standing water droplets are quite challenging to control, limiting the practical applications of the metasurface absorber. Another work that uses microfluidic reservoirs to regulate the droplet-based meta-molecules shows near-unity absorptions with incident frequencies ranging from 20 to 40 GHz, as shown in Fig. 10.3b [27]. The water droplets are well contained with the micro-reservoirs and can be dynamically controlled by the external pumping pressures. As a result, the absorption of the metasurface can be tuned via the changing of either external pumping pressures or the liquid material compositing the droplets. More importantly, the near-unity absorption remains when the metasurface is bent or attached to a curved substrate, indicating potential applications as smart coatings on irregular-shaped surfaces. A near-unity absorption from 5 to 32 GHz is later demonstrated by using the microfluidic metasurfaces, which are transparent in the visible frequency region and highly absorptive in the microwave region, as shown in Fig. 10.3c [28]. The water inside the microfluidic channels results in the refractive indexes matching between the microchannel and the PDMS surroundings, making the metasurface more transparent in the visible region due to the reduction of the scattering loss.

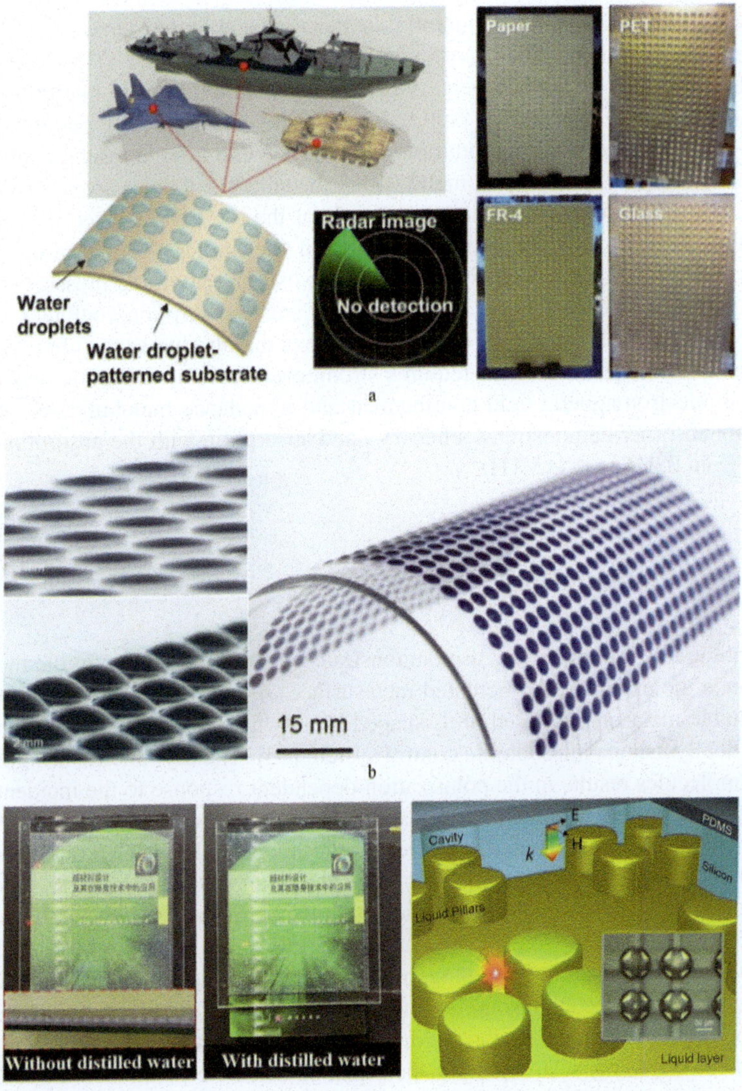

Fig. 10.3 Perfect absorber based on reconfigurable metasurfaces. **a**, **b** and **c** Droplet metasurfaces for broadband near-unity absorption working at the GHz region. Reproduced with permission [26] copyright from 2015 Springer Nature, with permission [27] copyright from 2017 Wiley–VCH, and with permission [28] copyright from 2018 AIP Publishing. **d** THz perfect absorber based on a multi-layered microfluidic metasurface with liquid metal. Reproduced with permission [29] copyright from 2017 AIP Publishing

The THz absorber based on microfluidic metasurfaces is also demonstrated using liquid–metal meta-molecules, as shown in Fig. 10.3d [29]. The THz metasurface absorber has three different layers. A PDMS layer on the top has microfluidic air channels linked to an external pump. A silicon layer periodically patterned with micro-holes is attached to the bottom PDMS layer with a large reservoir of liquid metal. The heights of the liquid metal within the silicon via holes are controlled by the pumping pressure within the air channels of the top PDMS layer. As a result, both the electrical and magnetic resonances of the liquid–metal meta-molecules can be controlled dynamically by changing the external pumping pressure. Liquid metal pillars replace the water droplets with a small absorption coefficient in the THz region. The absorption of the metasurface is mainly due to the ohmic loss of the liquid–metal pillars, which requires strong electrical and magnetic resonances for both electromagnetic field confinement and impedance matching. As a result, the THz absorber demonstrates a narrow band absorption with the absorption peak tuned from 0.246 to 0.415 THz.

10.4 Reconfigurable Polarizer

Reconfigurable polarizers are also demonstrated by using the microfluidic metasurfaces and the mechanically actuated metasurfaces. Figure 10.4a shows a function-switchable metasurface based on L-shaped liquid–metal meta-molecules regulated individually using a microfluidic control system [13]. The asymmetric design of the meta-molecules results in the polarization-dependent response to the incident electromagnetic waves, which can be tuned by changing the lengths of the two arms of the L-shaped meta-molecules. As a result, the microfluidic metasurface functions as a tunable polarizer working at the microwave frequency region. Another demonstration of polarization controllable microfluidic metasurface is shown in Fig. 10.4b, which is based on cross-shaped meta-molecules [30]. Here, the meta-molecules have independent responses to the orthogonally-polarized incidences, which supplies the metasurface with decoupled output wavefronts for x- and y-polarized incidences. The microfluidic metasurface shows multiple functionalities, which can be dynamically switched, including beam splitting, polarization selective beam steering, adaptive wavefront control, etc.

Figure 10.4c shows a metasurface polarizer with an MHz tuning speed, which is realized using the complementary structures of L-shaped meta-molecules fabricated on a GaAs membrane [31]. In experiment, the mechanically actuated metasurface membrane shows a dynamic control of the output polarization states and intensities with a 1.3 MHz modulation speed.

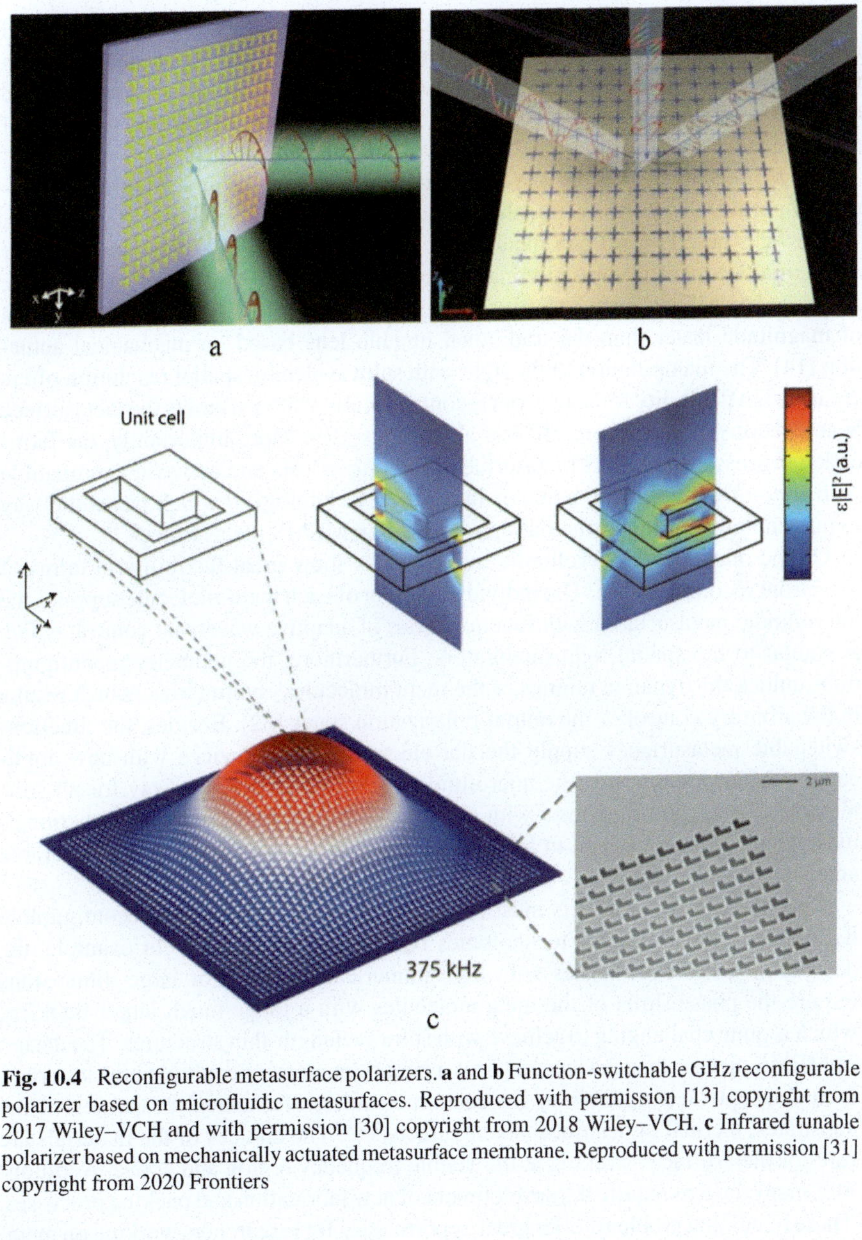

Fig. 10.4 Reconfigurable metasurface polarizers. **a** and **b** Function-switchable GHz reconfigurable polarizer based on microfluidic metasurfaces. Reproduced with permission [13] copyright from 2017 Wiley–VCH and with permission [30] copyright from 2018 Wiley–VCH. **c** Infrared tunable polarizer based on mechanically actuated metasurface membrane. Reproduced with permission [31] copyright from 2020 Frontiers

10.5 Summary and Outlook

Tunable and reconfigurable metasurfaces are developing diverse and promising applications due to their intriguing new physics. The MEMS and microfluidic metasurfaces have now shown substantial advantages in tunable flat optics devices, which break the design paradigms of the bulky devices based on refractive and reflective optics, resulting in improved performances and new functionalities. As a result, research efforts applied to the tunable and reconfigurable metasurfaces are dramatically increased, leading the tunable flat optical devices towards the route of commercialization.

The MEMS metasurfaces have been demonstrated with dynamic control of material properties and vast functionalities, including tunable lens, beam steering, etc. The devices based on MEMS metasurfaces have many merits compared with their counterparts consisting of traditional optical components. For example, the tunable lens based on MEMS metasurface can have a kHz modulation speed, which is orders of magnitude faster than the traditional tunable lens based on mechanical actuation [14]. The manipulation of the light with subwavelength spatial resolution offers great design flexibility for wavefront reconfiguration, which exempts the metasurface beam steering devices from off-axis aberrations [20]. More importantly, the fabrication processes of MEMS metasurfaces working at THz and above are compatible with state-of-the-art semiconductor fabrication technologies, which paves the way to mass manufacturing of the flat optics based on MEMS metasurfaces.

On the other hand, microfluidic metasurfaces show great flexibility in terms of wavefront reconfiguration. The individual tuning of each meta-molecule supplies the microfluidic metasurfaces with functionalities of adaptive wavefront control, which is similar to the spatial light modulators. Furthermore, the geometry reconfigurations unlock the dynamic tunings of the meta-molecules' symmetries, which results in the arbitrary control of the output polarization states [32]. Besides, the function-switchable metasurfaces supply the flat electromagnetic devices with new applications. More importantly, the microfluidic technology paves the way for all-soft-material metasurface absorbers with dynamic responses and switchable functionalities, which will surely lead to promising applications on smart coatings, wearable antennas, etc.

Nevertheless, many efforts must be made before the tunable and reconfigurable flat optics become matured technologies based on metasurfaces. For example, the designs of achromatic lenses with high numerical apertures or large dimensions require the phase shifts of the meta-molecules with a range much larger than 2π, which is quite challenging to achieve with subwavelength-thin structures. The design of MEMS metasurface devices requires a paradigm crossing multiple physics areas, including optical properties of the meta-molecules, mechanical properties of the substrates, thermal optical/mechanical effects, etc. The choices of the materials are rare for metasurfaces working at the visible frequency region and above. Additionally, many devices require the development of new fabrication and packing processes. These remaining problems offer great opportunities for researchers working on metasurfaces, optical MEMS, microfluidic chips, and optics designs, just to name a few.

References

1. Lee GJ, Choi C, Kim D-H, Song YM (2018) Bioinspired artificial eyes: optic components, digital cameras, and visual prostheses. Adv Func Mater 28(24):1705202. https://doi.org/10.1002/adfm.201705202
2. Bishara W, Sikora U, Mudanyali O, Su T-W, Yaglidere O, Luckhart S, Ozcan A (2011) Holographic pixel super-resolution in portable lensless on-chip microscopy using a fiber-optic array. Lab Chip 11(7):1276–1279. https://doi.org/10.1039/C0LC00684J
3. Kim I, Martins RJ, Jang J, Badloe T, Khadir S, Jung H-Y, Rho J (2021) Nanophotonics for light detection and ranging technology. Nat Nanotechnol 16(5):508–524. https://doi.org/10.1038/s41565-021-00895-3
4. Xie T, Guo S, Chen Z, Mukai D, Brenner M (2006) GRIN lens rod based probe for endoscopic spectral domain optical coherence tomography with fast dynamic focus tracking. Opt Express 14(8):3238–3246. https://doi.org/10.1364/OE.14.003238
5. Davis A, Kühnlenz F (2007) Optical design using fresnel lenses. Optik & Photonik 2(4):52–55. https://doi.org/10.1002/opph.201190287
6. Billings BH (1947) A tunable narrow-band optical filter. J Opt Soc Am 37(10):738–745. https://doi.org/10.1364/JOSA.37.000738
7. Okano F, Arai J, Okui M (2006) Amplified optical window for three-dimensional images. Optics Lett 31(12):1842–1844. https://doi.org/10.1364/OL.31.001842
8. Yu N, Capasso F (2014) Flat optics with designer metasurfaces. Nat Mater 13(2):139–150. https://doi.org/10.1038/nmat3839
9. Dorrah Ahmed H, Capasso F. Tunable structured light with flat optics. Science 376(6591):eabi6860. https://doi.org/10.1126/science.abi6860
10. Lin D, Fan P, Hasman E, Brongersma Mark L (2014) Dielectric gradient metasurface optical elements. Science 345(6194):298–302. https://doi.org/10.1126/science.1253213
11. Kamali SM, Arbabi E, Arbabi A, Faraon A (2018) A review of dielectric optical metasurfaces for wavefront control. Nanophotonics 7(6):1041–1068. https://doi.org/10.1515/nanoph-2017-0129
12. Wu J, Shen Z, Ge S, Chen B, Shen Z, Wang T, Wu P (2020) Liquid crystal programmable metasurface for terahertz beam steering. Appl Phys Lett 116(13):131104. https://doi.org/10.1063/1.5144858
13. Wu PC, Zhu W, Shen ZX, Chong PHJ, Ser W, Tsai DP, Liu A-Q (2017) Broadband wide-angle multifunctional polarization converter via liquid-metal-based metasurface. Adv Opt Mater 5(7):1600938. https://doi.org/10.1002/adom.201600938
14. Arbabi E, Arbabi A, Kamali SM, Horie Y, Faraji-Dana M, Faraon A (2018) MEMS-tunable dielectric metasurface lens. Nat Commun 9(1):812. https://doi.org/10.1038/s41467-018-03155-6
15. He S, Yang H, Jiang Y, Deng W, Zhu W (2019) Recent advances in MEMS metasurfaces and their applications on tunable lens. Micromachines 10(8). https://doi.org/10.3390/mi10080505
16. Wei Z, Cao Y, Su X, Gong Z, Long Y, Li H (2013) Highly efficient beam steering with a transparent metasurface. Opt Express 21(9):10739–10745. https://doi.org/10.1364/OE.21.010739
17. Fu X, Yang F, Liu C, Wu X, Cui TJ (2020) Terahertz beam steering technologies: from phased arrays to field-programmable metasurfaces. Adv Opt Mater 8(3):1900628. https://doi.org/10.1002/adom.201900628
18. Yan LB, Zhu WM, Wu PC, Cai H, Gu YD, Chin LK, Liu AQ (2017) Adaptable metasurface for dynamic anomalous reflection. Appl Phys Lett 110(20):201904. https://doi.org/10.1063/1.4983782
19. Meng C, Thrane Paul CV, Ding F, Gjessing J, Thomaschewski M, Wu C, Bozhevolnyi Sergey I. Dynamic piezoelectric MEMS-based optical metasurfaces. Sci Adv 7(26):eabg5639. https://doi.org/10.1126/sciadv.abg5639

20. Deng W, Zhu W, Shi Y, Liu Z, Zang G, Qin J, Zhu SJ (2022) Dynamic wavefront transformer based on a two-degree-of-freedom control system for 6-kHz mechanically actuated beam steering
21. Engelberg J, Levy U (2021) Achromatic flat lens performance limits. Optica 8(6):834–845. https://doi.org/10.1364/OPTICA.422843
22. Ee H-S, Agarwal R (2016) Tunable metasurface and flat optical zoom lens on a stretchable substrate. Nano Lett 16(4):2818–2823. https://doi.org/10.1021/acs.nanolett.6b00618
23. She A, Zhang S, Shian S, Clarke David R, Capasso F. Adaptive metalenses with simultaneous electrical control of focal length, astigmatism, and shift. Sci Adv 4(2):eaap9957. https://doi.org/10.1126/sciadv.aap9957
24. Zhu W, Song Q, Yan L, Zhang W, Wu P-C, Chin LK, Zheludev N (2015) A flat lens with tunable phase gradient by using random access reconfigurable metamaterial. Adv Mater 27(32):4739–4743. https://doi.org/10.1002/adma.201501943
25. Han Z, Colburn S, Majumdar A, Böhringer KF (2020) MEMS-actuated metasurface Alvarez lens. Microsyst Nanoeng 6(1):79. https://doi.org/10.1038/s41378-020-00190-6
26. Yoo YJ, Ju S, Park SY, Ju Kim Y, Bong J, Lim T, Lee Y (2015) Metamaterial absorber for electromagnetic waves in periodic water droplets. Scien Reports 5(1):14018. https://doi.org/10.1038/srep14018
27. Song Q, Zhang W, Wu PC, Zhu W, Shen ZX, Chong PHJ, Liu A-Q (2017) Water-resonator-based metasurface: an ultrabroadband and near-unity absorption. Adv Opt Mater 5(8):1601103. https://doi.org/10.1002/adom.201601103
28. Pang Y, Shen Y, Li Y, Wang J, Xu Z, Qu S (2018) Water-based metamaterial absorbers for optical transparency and broadband microwave absorption. J Appl Phys 123(15):155106. https://doi.org/10.1063/1.5023778
29. Song QH, Zhu WM, Wu PC, Zhang W, Wu QYS, Teng JH, Liu AQ (2017) Liquid-metal-based metasurface for terahertz absorption material: frequency-agile and wide-angle. APL Mater 5(6):066103. https://doi.org/10.1063/1.4985288
30. Yan L, Zhu W, Karim MF, Cai H, Gu AY, Shen Z, Liu AQ (2018) Arbitrary and independent polarization control in situ via a single metasurface. Adv Opt Mater 6(21):1800728. https://doi.org/10.1002/adom.201800728
31. Zanotto S, Colombano M, Navarro-Urrios D, Biasiol G, Sotomayor-Torres CM, Tredicucci A, Pitanti A (2020). Broadband dynamic polarization conversion in optomechanical metasurfaces. 7. https://doi.org/10.3389/fphy.2019.00231
32. Ren M-X, Wu W, Cai W, Pi B, Zhang X-Z, Xu J-J (2017) Reconfigurable metasurfaces that enable light polarization control by light. Light: Sci Appl 6(6):e16254–e16254. https://doi.org/10.1038/lsa.2016.254

The manufacturer's authorised representative in the EU is Springer Nature Customer Service Centre GmbH, Europaplatz 3, 69115 Heidelberg, Germany. If you have any concerns regarding our products, please contact ProductSafety@springernature.com

Printed and bound by CPI Group (UK) Ltd, Croydon, CR0 4YY
30/09/2026
02225336-0005